Subgrade Modeling and Models
in
Foundation Engineering

Front-cover graphic: This figure was published in U.S. Army Corps of Engineers Instruction Report ITL-94-6 for the computer program *CBEAMC*. This program was one of several created under the auspices of the Corps' Waterways Experiment Station (WES) Information Technology Laboratory (ITL) in Vicksburg, Mississippi as part of their Computer-Aided Structural Engineering (CASE) Project. Although this and other CASE software were developed for use by civilian civil engineers employed by the Corps as part of the Corps' extensive civilian mission, the later DOS-based microcomputer versions were unclassified and were once openly and easily available in the public domain for the 'price' of a blank 5¼" floppy disk.

CASE Project software was developed for specific, targeted applications that Corps engineers encountered routinely in their work. The *CBEAMC* program, which was developed under contract by then-Professor William P. Dawkins of Oklahoma State University in Stillwater, Oklahoma, was one of several programs for soil-structure interaction (SSI) analysis, in this case for a generalized 'true' beam-column with the option of various types of spring supports that could be used to crudely model a subgrade. The figure that is republished here depicts the sign conventions and generalized support and loading capabilities of the *CBEAMC* program.

Dr. Horvath used the *CBEAMC* program as an instructional tool at Manhattan College for many years as it was a relatively simple program that was legally available in the public domain for student download and use at no cost. Although the software was relatively crude by today's standards (it could only run on a personal computer in DOS mode using batch-type input and generated mostly text-format output), it was nonetheless successful for introducing both undergraduate and Master-level graduate students to basic concepts in SSI and subgrade modeling and models.

Rear-cover graphic: This figure is from a technical paper co-authored by Mr. Regis J. Colasanti, P.E., Member.ASCE and Dr. Horvath that appeared in the November 2010 issue of the scholarly journal *Practice Periodical on Structural Design and Construction*, an American Society of Civil Engineers publication. The figure depicts an isometric view (artificially expanded in the vertical direction for clarity) of the Horvath-Colasanti mechanical subgrade model as implemented by Mr. Colasanti in the computer program *ANSYS®* (Version 11.0). This was done specifically and primarily to illustrate how a state-of-art, multiple-parameter subgrade model that inherently incorporates the highly desirable feature of 'spring coupling' in its mathematical formulation could be implemented into commercially available structural analysis software and thus be usable to design professionals in routine practice.

The input parameters for the Horvath-Colasanti Model were quantified by using the hybrid subgrade model concept and concomitant coefficient matching with the Reissner Simplified Continuum subgrade model. The Horvath-Colasanti/Reissner Model is the most advanced hybrid subgrade model that has been identified and fully developed to date. This includes theoretical development of subgrade boundary conditions so that the subgrade beyond the edge of the loaded area (mat in this case) does not have to be modeled explicitly.

Subgrade Modeling and Models in Foundation Engineering

A Scholarly Monograph
by

John S. Horvath, Ph.D., P.E., Life Member.ASCE

Consulting Professional Engineer
d/b/a
John S. Horvath Consulting Engineer
Scarsdale, New York, U.S.A.

and

Professor of Civil Engineering (retired)
Manhattan College
School of Engineering
Civil and Environmental Engineering Department
Bronx, New York, U.S.A.

Published
by
John S. Horvath Consulting Engineer
Scarsdale, New York, U.S.A

Subgrade Modeling and Models in Foundation Engineering
by John S. Horvath

This document is a scholarly monograph written solely for the purpose of dissemination of technical knowledge and information for consideration by licensed design professionals and academicians educated in the field of civil engineering. As such, it is not a textbook, design manual, or standard, and should not be used or referenced as such. Furthermore, no liability is expressed or implied as to the suitability or accuracy of any information presented herein for any specific application in engineered construction or academic research. Use of any of the methods presented herein remains solely the liability of the licensed design professional using these methods.

Published August 2018
by
John S. Horvath Consulting Engineer
Scarsdale, New York, U.S.A.
linkedin.com/in/jshce

Library of Congress Control Number: 2018908787

ISBN-13: 978-1-7320953-1-1

Contents

This page intentionally left blank.

Preface

This monograph marks a return to the roots of my very first efforts in scholarly research that involved subgrade models in general and their applications to mat (raft) foundations in particular. I began this research in 1973 in a very inauspicious way. I was still relatively new to my first permanent professional employment (begun in June 1972, with the Port Authority of New York and New Jersey) when I was tasked one day by my supervisor to come up with a value of the 'soil spring constant' for some long-forgotten project in the New Jersey-New York Port District. While attempting to fulfill this task by looking at all the usual textbooks of the era, I was struck by how subjective and overall 'black box' mysterious this process was. I was so frustrated intellectually (yet simultaneously intrigued and motivated to do better) by this supposedly simple, trivial exercise that I decided that same year to further my post-graduate education on a part-time basis by enrolling at the Polytechnic Institute of New York (PINY)[1] and conduct doctoral-level research into the topic of subgrade modeling and models. This effort culminated in my receiving a Doctor of Philosophy degree from PINY six years later, in 1979.

Although my research interests since the 1970s have gone off into several diverse topical areas (cellular geosynthetics/geofoam and tapered piles are at the top of the list), I neither forgot nor gave up my interest in subgrade modeling and models. To a large extent, this was due to the fact that I was not wholly satisfied with the outcomes of my doctoral research. While my work satisfied the statutory degree requirements of PINY, I was personally dissatisfied with not being able to adequately implement my research outcomes in commercially available computer software that was accessible to design professionals in everyday practice (my research interests have always been practice-oriented in nature). It was, to me, an example of the classic case of satisfying the letter of the law but not the spirit of the law. In this case, this was due to the state of knowledge being beyond the then state of practice in terms of computational tools available to practicing engineers (note that the 1970s were still very much the punch card/mainframe computer era, at least for civil engineering work).

Fortunately, I eventually realized my goal of matching knowledge and practice decades later in the early years of the 21st century with the crucial and outstanding professional assistance and collaboration of Mr. Regis J. Colasanti, P.E., Member.ASCE, a

---

[1] At the time, PINY was a very recent (1973), forced (by the State of New York which had direct financial investments and concomitant interest in the matter) amalgamation of the Polytechnic Institute of Brooklyn (PIB) and the original engineering school of New York University (NYU). The latter was located some distance from NYU's main Washington Square campus in Manhattan, in the University Heights section of the Bronx. All facilities of the newly formed PINY were consolidated at PIB's main campus in downtown Brooklyn which at that time consisted primarily of a former razor blade factory. Consequently, in reality it was not a merger of equals but an acquisition of NYU's engineering and related technologies program by PIB. The underlying cause and genesis of this amalgamation was that both schools had invested heavily in aerospace-related education and research (PIB even had a satellite campus located near Grumman on Long Island) and both suffered from the downturn in aerospace R&D funding once the 'space race' to the Moon was over. Circa-1970s financial problems of New York City as well as shifting demographics in the Bronx also played causative roles in this amalgamation. PINY went on to become Polytechnic University (in 1985) and subsequently, through an interesting example of what might be called circular fate, was acquired by NYU in the early years of the 21st century and became the NYU Tandon School of Engineering in 2015. It remains on a much-expanded and -improved campus in downtown Brooklyn.

structural engineering practitioner whose intellectual abilities, practical experience, and computer skills with structural engineering software provided the structures-oriented intellectual ingredient that I had been seeking since 1979 to complement my geotechnical knowledge. I also acknowledge with gratitude Mr. Colasanti's employer at the time (URS Energy and Construction Division, the former Washington Group International) for graciously providing him both access to state-of-art computer software (*ANSYS®*) as well as release time to perform structural analyses both as I requested and as he suggested to me. I note that Mr. Colasanti is one of the very few people with whom I have chosen to co-author published work in my professional career.

An additional benefit of continuing to ponder the general subject of subgrade modeling and models for the past four decades is that it has allowed me to continuously refine and crystallize my own thoughts and insight into the subject which were still developing and evolving when I completed my doctoral work in 1979. In this regard, early on (circa 1990) I benefitted in particular from conversations with a few individuals, most notably Samson S. C. 'Sam' Liao, Ph.D., P.E. and the late Prof. Nicholas F. Morris, Ph.D., P.E. of NYU and later Manhattan College. I also acknowledge and thank Dr. Liao's employer at the time (Parsons Brinckerhoff, Inc.) for graciously allowing me access to review the outcomes of internally funded R&D work related to subgrade modeling and models that was performed by him.

My ongoing thought process concerning subgrade modeling and models has now reached the point where I feel that a scholarly monograph on the subject is warranted. The single most important goal of this monograph is to illustrate in simple, basic terms why the profession needs subgrade models in the first place as I believe this goes a long way toward correctly understanding what subgrade models and, especially, the model parameters are and are not. If nothing else, my intention is to illustrate once and for all that this mysterious parameter that is named the coefficient of subgrade reaction but often referred to colloquially as the 'soil spring constant' or some variant thereof is not, has never been, and never will be an inherent soil/subgrade material property. It is, in reality, an arbitrarily defined parameter that varies in magnitude from point to point in a soil-structure interaction problem and even at a given point will vary with changes in loading to the overall system being analyzed.

In closing, I fully realize that we live in an age where finite-element and similar numerical analyses of continuums are performed routinely and increasingly for project applications that once relied on simpler analytical strategies that incorporated a subgrade model. However, I believe that the need to perform simpler analyses using subgrade models still exists in both routine practice and academic research. Consequently, a considerable portion of this monograph is devoted to practice-oriented considerations related to subgrade modeling and models that will hopefully find their way into use by both design professionals and academic researchers alike.

As a final comment, several brands of commercially available software are mentioned by name in this monograph. This is solely for informational purposes and does not in any way constitute or imply a recommendation by me. And to leave no stone unturned or question unanswered, there has never been any external financial support of any kind for any of my research and writing related to the topics covered in this monograph.

John S. Horvath, Ph.D., P.E., Life Member.ASCE
Scarsdale, New York, U.S.A.
August 2018

Chapter 1

Introduction

1.1 PROLOGUE

1.1.1 Overview

The overall topic of this monograph...subgrade modeling and models in foundation engineering...is one that is already broadly familiar to both practitioners and academicians alike. Subgrade models and, especially, opinions on how to evaluate subgrade model parameters have been the subject of countless scholarly publications such as technical papers, research reports, and design manuals going back to seminal work in Europe the 19th century[2].

However, in the writer's experience this knowledge is generally incomplete and superficial. More significantly, this knowledge often contains significant and important errors of fact. Furthermore, many substantive issues, including basic subjects such as how and why subgrade modeling came to be needed in the first place and how subgrade-modeling needs have changed significantly over time, are neither widely nor adequately known.

This overall unsatisfactory state of knowledge may be due, at least in part, to the fact that insightful, theoretically rigorous knowledge about subgrade models did not appear in published work until almost 100 years after the Fuss-Winkler Hypothesis was published. This fact of theoretical evolution allowed the simple, abstract Fuss-Winkler Hypothesis, which is totally lacking in theoretical basis and rigor, to become well established and flourish to the point where it is typically all that many civil engineers know about subgrade models.

This is conceptually similar to the evolution and persistence of knowledge related to pile-driving dynamics where simple concepts based on energy transfer and rigid-body dynamics as epitomized by the *Engineering News* formula...concepts that are now known to be simply incorrect physically...appeared a century before the one-dimensional (1-D) wave-equation concept became not only widely known but computationally solvable using the digital computer. If there is nothing else that the writer has learned after more than 46 years of professional experience, it is that once a simple idea establishes a technological beachhead and takes root in routine practice it becomes very difficult to dislodge it, even when compelling evidence is presented to indicate that the simple concept is not only incorrect but possibly unconservative and thus potentially unsafe as well.

---

[2] The mathematical abstraction called *Winkler's Hypothesis* that is for all practical purposes the original subgrade model is credited by most researchers in the English-language literature (e.g. Hetényi 1946, Rhines 1965, Nair and Chang 1973) to the German civil engineer Emil Winkler who reportedly presented his concept in a book titled *Die Lehre von der Elastizität und Festigkeit* that was published in 1867. However, in the Russian-language literature Vlasov and Leont'ev (1960) claim that the concept was actually first proposed decades earlier, in 1801, by the Swiss mathematician Nicolas Fuss. Fuss spent his entire adult life in Imperial Russia where early on he was a protégé of Leonhard Euler. Euler was another Swiss expatriate in Imperial Russia and his work related to beams is noted in this monograph. Given the fact that the writer cannot independently assess these competing claims of invention and in the interest of objectivity until proven otherwise, the term *Fuss-Winkler Hypothesis* is used in this monograph in lieu of Winkler's Hypothesis as the writer used in earlier publications on the subject.

A further issue that complicates the subject of subgrade modeling and models is terminology and notation. Compared to many other disciplines in civil engineering, terminological and notational standardization has always been and still is lacking in geotechnical and foundation engineering. This has only become worse over time as access to material published globally by a more-diverse authorship has become routine.

What makes this issue particularly severe and problematic with subgrade modeling and models is that the subject draws on an unusually diverse body of work in not only geotechnical and foundation engineering but structural engineering and applied/engineering mechanics as well. Furthermore, this collective body of work has been published in many countries, languages, and alphabets over a span of approximately 200 years. As a result, one routinely finds that the same parameter has multiple names and notations in different bodies of work. Even worse is that the same name and notation can be used for different parameters depending on the preferences of the author. All of this adds to the confusion and misunderstanding that seems to be particularly severe with the topic of subgrade modeling and models compared to others in geotechnical and foundation engineering.

Thus, the writer's assessment of the state of practice with respect to the knowledge base of subgrade modeling and models is that although the basic concept may be well known there remains an unfulfilled need to first and foremost place subgrade modeling in its proper, complete theoretical and historical context. This will hopefully lead to more-rational selection and parameter assessment of subgrade models in foundation engineering applications.

Finally, before outlining the scope and content of this monograph in detail, and to provide background and context for the topics covered in this monograph, it is useful to discuss two topics that are fundamental to the subjects to be covered: *foundation engineering* and *soil-structure interaction*. While these topics may seem to be intuitive and obvious, the writer has found that there are subtle aspects of each that should be clarified and understood in order to better serve the goals of this monograph.

1.1.2 Foundations and Foundation Engineering

Foundation engineering can be defined as the professional engineering efforts that go into crafting a structural element or group of elements called a *foundation* whose primary, if not sole, purpose or function is to transfer the relatively concentrated load(s) from a *superstructure* to the ground. Nowadays, it is expected that this load transfer be done in a controlled, forecastable[3] manner so that the superstructure functions (performs) as intended for its design life.

---

[3] Foundation engineers routinely use a variety of terms (e.g. predict/prediction, estimate/estimation) to define the expected performance of some design. The term 'prediction' is perhaps the most widely used as can be seen in the many 'prediction symposiums' that have been conducted over the years. However, the writer feels that more attention should be paid to the terminology used as this aspect of foundation engineering is one that can interface with the non-engineering public at times, especially when things 'go bad' on a high-profile project such as the Millennium Tower in San Francisco that was completed earlier in the 21st century. As such, foundation engineers should be sensitive to how the public hears, i.e. interprets, engineering terminology. In particular, the writer posits that the terms 'predict' and 'prediction' convey a degree of near certainty or confidence. Consequently, it is suggested that the term 'forecast' is better suited for use. The public is familiar with this term from weather forecasts and understands that a forecast conveys an estimated behavior that is reasoned and has scientific basis for high probability yet is subject to some uncertainty due to the vagaries of nature. Thus, the public understands (or at least should understand) that a forecast is not a statement of near certainty and thus allows (or should allow) some leeway **[continued on following page]**

Note that within this intentionally broad definition of a foundation being a load-transfer structural element between superstructure and subgrade the writer includes applications such as slabs-on-grade, pavements, and railway track systems. This is because the concepts of subgrade modeling and models presented in this monograph are equally applicable to these common civil engineering applications as well.

As an aside, there are situations where the distinction between foundation and superstructure cannot easily be made. Examples include gravity retaining walls and masonry load-bearing walls of older buildings. In each case, the wall itself is so wide, at least at its base, that it acts as its own continuous- (strip-) footing foundation.

It can be posited that foundation design...at least as an art developed through human trial-and-error and experience as opposed to an applied science implemented by educated design professionals through a rational, mathematized process...is the oldest aspect of civil engineering. There is archaeological evidence of multiple European lake-dwelling cultures that inhabited structures supported on wooden columns sunk into the ground...what would today be called *driven timber piles*...as far back as 5000 years BCE[4].

When civil engineering began to emerge as a recognized profession in the 19th century, foundation engineering was initially viewed as a component or subset of structural engineering as opposed to being an entity in and of itself. That foundation design would fall under the purview of structural engineering is logical given the fact that foundation elements have always been, first and foremost, structural elements that are literally extensions of the superstructure. Even though modern soil mechanics began to evolve in the 1920s and 1930s, the recognition of geotechnical engineering as being a separate, essential component of foundation engineering was slow to occur in many countries, including the U.S. It was well into the 1960s before geotechnical coursework was required of all civil engineering undergraduates in the U.S. This is a not-insignificant fact given that a Bachelor degree was the terminal degree for most civil engineering practitioners at the time. Early in the writer's career (circa 1970s), older design professionals were routinely encountered who still felt that foundation engineering was something adequately handled solely by structural engineers and that hiring a geotechnical engineering consultant for input into foundation selection and design was an unnecessary expense. Even to the present, some state departments of transportation in the U.S. are structured so that the geotechnical group is only responsible for tasks such as standalone embankments and slope stability along rights-of-way. Foundations for structures such as bridges are solely within the purview of the bridge engineering group that is populated and managed by structural engineers. The hoary chestnut of "When in confusion, drive piles in profusion"...indicating a foundation selection process based more on technical ignorance and historical usage ("We've always done it this way...") than technical need and science...was and still is fact, not fantasy, on more than a few projects in all-too-many state DOTs.

It is relevant to note that even today in a 'perfect' world that is defined by the ever-more-specialized nature of civil engineering, foundation engineering remains an enduring partnership between geotechnical and structural engineering specialists as each has significant contributions to make to the final, project-specific outcome. This point is emphasized here as to a very significant extent the diverse requirements of geotechnical and structural engineers in foundation design have dominated and continue to dominate the subject of this monograph, subgrade modeling and models in foundation engineering.

---

if actual outcomes deviate from those forecast. In summary, the writer feels that foundation engineering outcomes should be presented as 'forecasts', not 'predictions' or 'estimates', and will so use the term 'forecast' throughout this monograph.

[4] *Before Common Era*, used as an alternative to *BC*.

4

Specifically, as will be seen, the specific requirements of structural engineers in foundation design in routine practice have dictated the need for subgrade modeling and the concomitant development of subgrade models by geotechnical engineers. This is a relatively uncommon situation in modern civil engineering, where the technical needs of one specialty or discipline completely dictates the outcomes provided by another discipline.

1.1.3 Soil-Structure Interaction

Problems in foundation engineering are the major component of a diverse group of civil engineering applications that involve some type of structural element in direct contact with the ground[5]. When forces are either applied externally to the structural element or develop internally within the ground, both problem components (structural element and ground) must displace in a compatible manner because of their intimate physical contact. Therefore, these are broadly referred to as *soil-structure interaction* (SSI) problems, even when the ground (which is also referred to as the *subgrade*) in contact with the structural element is composed of rock or non-earth material.

Strictly speaking, SSI is present to some degree in every foundation application. However, historically and most commonly, civil engineers limit the SSI terminology to applications where the foundation element exhibits significant flexibility and deformation under service loads so that displacement (settlement in most cases) of the foundation element cannot be adequately characterized by a single value. Rather, displacements are varying continuously along the foundation-subgrade interface. Typical situations where this occurs involve mat (raft) foundations, slabs-on-grade, pavements, railway track systems, and laterally loaded deep foundations. On the other hand, spread footings are generally not considered SSI applications as single-value settlements are adequate to define their behavior.

Note that some civil engineers (especially in older publications) have adopted a much narrower perspective and limited the SSI terminology to problems involving seismic loading only. The writer rejects this narrow interpretation as being too limiting. Therefore, this monograph will follow the more-common usage of SSI to cover all types of loading in problems where flexibility and deformation of the foundation element is important in problem solution for both the foundation element and superstructure that it supports.

There are two aspects of SSI that are particularly relevant to this monograph:

- The need for considering SSI in many application categories is what led to subgrade modeling and the concomitant development of subgrade models.

- If and how SSI is considered in an analysis has always been a tradeoff between technical need and the computational tools available. Because these computational tools change over time, this means that how the linked concepts of SSI, subgrade modeling, and subgrade models relate to each other is never technologically static but always in a state of flux.

Note, however, that the current state of practice in foundation engineering still includes many instances where SSI is legitimately neglected (it was almost always necessary to do so in the days before digital computers) so that the foundation element and subgrade are analyzed independently of each other. This is done primarily for analytical simplicity

---

[5] Non-foundation applications in this category are primarily flexible earth-retaining structures such as sheet-pile walls and bulkheads.

because SSI analyses are inherently statically indeterminate which means that in addition to satisfying force and moment equilibrium displacements must be considered explicitly to rigorously solve any SSI problem. While indeterminate analyses are easier than ever to perform given the current state of computational hardware and software available to design professionals in routine practice, such analyses are still relatively complex and time consuming, especially for combined superstructure-foundation-subgrade systems. Thus, the pragmatic incentive to simplify the analyses and not consider SSI for simple, routine applications is still very strong in practice.

In most cases, neglecting SSI is quite reasonable and justified by experience based on satisfactory performance of the resulting design. This typically involves applications where two assumptions can reasonably be made:

- the foundation is relatively rigid compared to the subgrade so that displacement (typically settlement) of the foundation can reasonably be assumed to be single-valued and

- differential settlement of the superstructure due to varying settlement of discrete foundation elements and the effect of this differential settlement on forces and moments within the superstructure can be neglected.

A common example where these conditions are routinely met in practice is footing foundations supporting a low-rise building.

Unfortunately, there are also cases where neglecting SSI is no longer justified based on the current state of knowledge and computational technology yet is still done for any number of non-technical reasons (tradition, ignorance, mental inertia). This can lead to designs that perform unsatisfactorily. Therefore, it is important for a civil engineer to understand that SSI is always present for all types of subgrades even if the engineer chooses to neglect it either explicitly or implicitly by virtue of an analytical methodology that is employed. If SSI is neglected, there should always be sound theoretical and practical bases for so doing.

1.2 OBJECTIVES AND SCOPE

The primary objectives of this monograph are to explain:

- the genesis of subgrade modeling for SSI analyses;

- the tectonic paradigm shift reflected in the transition of *subgrade reaction*[6] from a calculated result of secondary importance to an essential, primary problem variable that must always be defined in some manner as problem input prior to the start of an SSI analysis; and

- the subgrade models that have evolved from this need to be able to mathematically define subgrade reaction beforehand.

While these subjects are not unique to this monograph, what is intended to be the defining uniqueness of this monograph is the manner of presentation that will hopefully provide the

---

[6] This term is defined in Chapter 2.

insight into the overall subject that has been missing in the published literature to date, including earlier published works by the writer.

Because this document is intended to be a resource document for both design professionals in practice as well as academic instructors and researchers, also presented in this monograph are practice-oriented discussions of parameter assessment for subgrade models as well as several case history applications using advanced, state-of-art subgrade models.

One element that has been retained from the writer's early published works on the subject is the use of a mat (raft) foundation as an exemplar of a classic SSI problem in foundation engineering. There are a number of compelling reasons for doing so. Consequently and to provide continuity, the use of a mat foundation is carried through this entire monograph, from illustrating basic elements of subgrade modeling, to the use of subgrade models, and for case history applications comparing observed results and forecast outcomes using the most advanced subgrade model that has been implemented into practice to date. However, the material presented is sufficiently broad and generic so that it can be extended to any of the foundation-related SSI applications noted previously (e.g. slabs-on-grade) and even non-foundation SSI applications such as sheet-pile walls as well.

1.3 CONTENT AND ORGANIZATION

To meet the desired primary objectives of providing a unique, improved insight into fundamental aspects of subgrade modeling and models, the presentation of material in this monograph is structured differently from that used historically, including in earlier published works by the writer. Specifically, the overall structure of this monograph will follow a central theme of theoretical logic to show how subgrade models are pragmatic simplifications of an 'exact' solution to a problem involving SSI. This will be done in a manner so that the simplifications inherent in any subgrade model are clearly apparent.

This approach deviates from the presentation structure normally seen in published work, including earlier work by the writer (Horvath 1979, 1988d, 1989c, 2002), where a temporal format is used to follow subgrade models in order of their appearance in the published literature. This is a useful approach for many geotechnical engineering topics where knowledge builds on itself in a logical manner over time. However, after working with subgrade models for more than four decades, the writer has concluded that the subject could benefit from a different presentation approach.

The remaining sections of this monograph are as follows:

- Chapter 2 defines structural engineering concepts that are essential, integral elements of SSI applications;

- Chapter 3 outlines the genesis of subgrade modeling and models and, most importantly, explains the requirements and limitations placed on subgrade model development over time;

- Chapter 4 outlines the history of subgrade models and integrates this history in a way to make subgrade models, especially advanced multi-parameter models, more user friendly;

- Chapter 5 provides insight into practical aspects of parameter assessment for subgrade models;

- <u>Chapter 6</u> presents results of several case histories that were evaluated using outcomes presented in Chapters 4 and 5;

- <u>Appendix A</u> discusses the question of whether or not to include the weight of the foundation element that is bearing on or in the ground in an SSI analysis;

- <u>Appendix B</u> discusses what the writer calls the *Conventional Method of Static Equilibrium* that is not, strictly speaking, a subgrade model but has historically functioned as one and thus played an outsized role in some SSI applications such as mat foundations until the relatively recent past so merits some discussion to clearly illustrate its place in the evolution of subgrade modeling and models;

- <u>Appendix C</u> illustrates the theoretical developments of the parameters for the *Horvath-Colasanti/Reissner hybrid subgrade model* as well as boundary conditions for this model when used for applications such as mat foundations and slabs on grade;

- <u>Appendix D</u> illustrates the theoretical elements of the writer's extension, called the *MTH Method*, of the Charles *MT Method* for determining the equivalent depth to a rigid base and related elastic parameters for analytical purposes relevant to the hybrid subgrade models discussed in this monograph; and

- <u>References</u> cited in this monograph as well as a limited bibliography of additional, uncited publications of relevance to the topics covered in this monograph.

1.4 SYSTEMS OF UNITS

Almost all of the writer's publications (which go back to 1977) have expressed dimensioned parameters using both the *Imperial* (a.k.a. *U.S Customary* or *English*) system of units and the *Système International d'Unités* (SI) version of metric units that is considered the preferred standard for civil engineering. Furthermore, a given document typically used one system as the primary and the other as the secondary consistently throughout the document. Which system was which depended on either a mandated requirement of the publishing agency (e.g. American Society of Civil Engineers) or the writer's preference based on the anticipated audience or perceived state of practice at the time the document was written.

This monograph deviates from this practice to the extent that there is no primary system of units although dual units are given in almost all cases. While the SI system is primary in most cases (for some years now, the writer has performed research solely using SI units to enhance the international usability of published results), for case histories and other situations where the original work was published using Imperial units alone or primary this system is retained as primary with SI equivalents secondary. This is because in those cases the Imperial-unit dimensions are 'exact' and should thus be expressed as such.

However, one issue that is consistent throughout this monograph is that a decimal point (.) is used to indicate a fractional value of a number, including those with SI units in some cases (SI rules deprecate this practice but, in some cases, it was judged to be pragmatic). Also, with Imperial units a comma (,) is used to separate groups of three integer values (this is avoided with SI units to be consistent with SI rules). This usage is highlighted as in some parts of the world a comma is used instead of a decimal point to indicate a fractional value of a number.

This page intentionally left blank.

Chapter 2

Structural Engineering Concepts

2.1 INTRODUCTION AND OVERVIEW

As noted in the Chapter 1, structural engineering is an integral and at times dominant part of foundation engineering. This is especially true of the SSI-related topics that are the focus of this monograph. Therefore, before getting into the primary and more-geotechnical content related to subgrade modeling and models, it is useful to present a primer on a variety of structural engineering concepts that the writer has found are essential to the contents of this monograph. This is done to:

- efficiently summarize in one place concepts in structural mechanics that are relevant to SSI problems;

- standardize the structural-mechanics terminology and notation that is used throughout this monograph; and

- highlight, and in some cases clarify, behavioral aspects of structural elements that are used in various aspects of subgrade modeling and models.

2.2 NOTATION

2.2.1 Overview

There are consistent, significant differences in the notation used by geotechnical and structural engineers, at least in U.S. practice. Because of the close connection and working relationship between these two civil engineering disciplines in foundation engineering in general and SSI analyses in particular, and especially because one major element of this monograph on how to implement geotechnical subgrade models within the context of commercially available structural analysis software, it is important to discuss these differences in order to minimize the potential for misunderstanding and error.

The differences can be divided into two categories:

- coordinate axes and associated displacements, and

- forces and stresses.

2.2.2 Coordinate Axes and Displacements

Both geotechnical and structural engineers define positive normal displacements $(+u,+v,+w)$ as displacements in the positive direction of the relevant coordinate axis $(x,y,z$ respectively), with the axes in each case defined using the 'right-hand rule'. The differences between the sign conventions used by these two civil engineering disciplines are:

- the spatial orientation of the positive sense of the vertical axis and

- the specific axis labels.

The most common and consistent versions of geotechnical and structural sign conventions that the writer has seen used in U.S. practice are shown in Figure 2.1 for general three-dimensional (3-D) conditions. Also shown in this figure is that in geotechnical applications the Greek letter ρ is often used as the settlement variable instead of w.

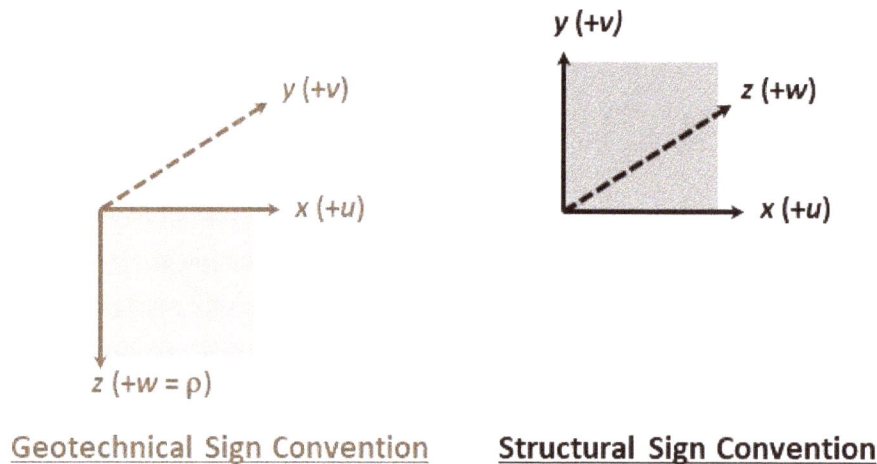

Geotechnical Sign Convention   **Structural Sign Convention**

Figure 2.1. Geotechnical vs. Structural Axes Orientations and Displacement Sign Conventions.

Note that sign conventions other these two can readily be found in the published literature related to SSI applications. For example, the seminal scholarly monograph on SSI by Hetényi (1946) uses what is basically the geotechnical sign convention shown in Figure 2.1 but with y for the downward-positive vertical axis in lieu of z.

In addition, some authors use the same variable notation for both the axis and normal displacement associated with that axis as opposed to using two distinct variables as shown in Figure 2.1. Again, with reference to Hetényi (1946), the notation y is used for both the axis and normal displacement (settlement in this case) associated with that axis.

All in all, these variations in even basic notational usage can present a confusing picture and emphasizes the fact that even seemingly trivial technical issues should always be clearly identified beforehand.

As will be seen, much of material presented in this monograph is for two-dimensional (2-D) conditions, solely for simplicity of illustration and discussion without any loss of generality for 3-D applications. This means that the geotechnical sign convention reduces to the x-z plane whereas the structural sign convention reduces to the x-y plane, both as shown by the shaded areas in Figure 2.1.

One final comment with regard to displacements is that in geotechnical analyses it is ubiquitous to refer to vertical displacements, w, as either *settlement* (if downward in the $+z$ direction) or *heave* or *rebound* (if upward in the $-z$ direction). Furthermore, unique notation is generally used for settlement (as noted above, ρ is most common and used in this monograph) although there does not appear to be any consensus notation for heave/rebound.

That having been said, much of the presentation in this monograph involves algebraic expressions for subgrade models and structural behavior. Consequently, the more-general term 'displacement' and notation 'w' will generally be used for settlement and heave/rebound even in a geotechnical context.

2.2.3 Forces and Stresses

With regard to forces (and, by extension, normal stresses), structural engineers generally follow traditional solid mechanics convention and define tension as positive. On the other hand, geotechnical engineers traditionally define compression as positive. This is because particulate materials such as soil have no tensile strength so tensile forces and stresses are relatively uncommon in geotechnical problems. Defining compression as positive is simply pragmatic as it removes the need to continually use negative signs.

2.2.4 Monograph Usage

Because the focus of this monograph is on the geotechnical aspects of subgrade modeling, the geotechnical sign conventions outlined in the preceding sections is used in most cases, including when nominally structural aspects of the problem are investigated. Consequently, topics covered in this chapter such as the behavior of beams are illustrated using geotechnical, not structural, sign conventions. This is done solely to unify the theoretical structural and geotechnical presentations in this monograph and does not detract from the generality of the outcomes.

Exceptions to this use of geotechnical sign conventions are made when using structural or both sign conventions is important for historical purposes. This is true especially in Chapter 3 where it is relevant to show the interaction between structural and geotechnical engineers over the years in SSI applications, especially mat foundations.

2.3 FLEXURAL BEHAVIOR

2.3.1 Overview

Overall, flexural (bending) behavior has the greatest influence on the structural modeling aspects of SSI applications. This applies to when a structural element physically exists in a problem as the foundation element (e.g. mat foundation) as well as when a structural element is used in a fictitious, conceptual context to represent a component of a subgrade model. Thus, it is important to discuss various issues that relate to flexural behavior.

These issues can be broadly divided into two categories:

- those related to theoretical structural-mechanics issues that exist independent of the specific material used for the structural element although the type of material behavior assumed (linear elastic, elastoplastic, etc.) usually must be specified and

- those related to a specific material used.

The following presentation is organized along these lines and each of these factors should be considered in practice.

For simplicity in presenting and discussing basic concepts, the following presentation focuses on the basic 2-D problem of uniaxial bending of a beam with the following generic properties:

- homogeneous, isotropic composition of a linear-elastic material with a Young's modulus, E;

- prismatic cross-section of unit width; and

- moment of inertia, I.

However, the concepts discussed and conclusions reached are completely general so can be extended to the more-general 3-D problem involving biaxial bending of structural elements such as plates. Although some general comments regarding plate behavior are included in this discussion for the sake of completeness, the specifics of this extension are not covered in this monograph as they are dealt with in other sources.

2.3.2 Structural-Mechanics Issues

2.3.2.1 Overview

The load-deflection behavior of a beam composed of linear-elastic material and subjected to flexure is one of the most basic concepts in structural mechanics that is taught to all civil engineers as part of their undergraduate education. However, there are numerous assumptions that can be made in the process of developing a mathematical model and solution to this elementary problem. Each of these assumptions impacts the theoretical accuracy and concomitant calculated results obtained using the developed solution.

The nuances of the different versions of the basic beam-flexure problem are important to the SSI applications of interest in this monograph. For example, the design of many mat-supported structures is governed by differential settlements of the superstructure which, in turn, are influenced by the flexural behavior of the mat. Thus, anything that influences the flexural stiffness and concomitant flexural behavior of a mat should be clearly understood by the design professionals involved so that the analytical capabilities of commercially available structural analysis software are utilized appropriately for the needs of a given project.

The traditional mathematical format for developing and portraying the load-displacement behavior of a beam in uniaxial bending is an ordinary-differential equation. Such equations are the direct outcome of applying the fundamental concepts of static equilibrium of forces and moments to a free-body diagram of forces developed for some assumed physical model of the flexural response of a beam under various boundary conditions. This traditional differential-equation format is used extensively in this monograph both for its historical significance as well as to illustrate differences in outcomes from different versions of the basic beam-flexure problem.

The decades-long development and use of the digital computer for structural analysis based on the *matrix method*[7] provides an alternative format of mathematical expression for the basic beam-flexure problem. The basic form for beam flexure using matrix notation is:

---

[7] The matrix method is the conceptual twin of the well-known *finite-element method* (FEM) that is used extensively for many geotechnical and foundation engineering problems.

$$[S]\{d\} = \{q\} \tag{2.1}$$

where:
- $[S]$ = stiffness matrix,
- $\{d\}$ = displacement vector, and
- $\{q\}$ = load (force) vector.

This alternative format is used in this monograph as well because matrix notation is particularly insightful into the subtle variations in beam behavior that result from the different formulations and solutions noted above. The particular utility of Equation 2.1 is that most of the variations in beam behavior discussed subsequently can be expressed solely as variations in the contents of the stiffness matrix, $[S]$. The remaining variations require at most the additional of another vector to Equation 2.1. Thus, matrix notation provides a concise and easily visualized methodology for understanding all the variations in the formulation of the overall beam-flexure problem.

2.3.2.2 Traditional Solution (Euler-Bernoulli Beam)

Figure 2.2 illustrates the basic components of a beam subjected to transverse applied loads $q(x)$. Note that the depiction of $q(x)$ in this and subsequent figures is simplistic as $q(x)$ is not necessarily uniform in magnitude or even continuous along the length of the beam. Details regarding the beam supports (roller, hinged, fixed, or other) do not qualitatively affect the current discussion so are omitted for clarity.

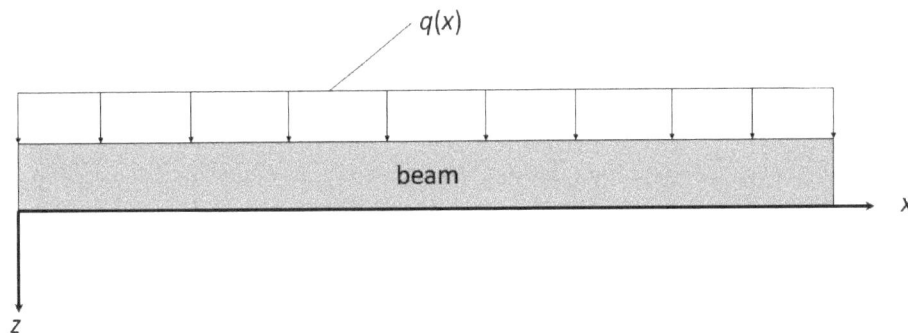

Figure 2.2. Beam with Transverse Applied Load.

In developing the traditional solution for the z-axis displacement along the length of the beam, $w(x)$, of what it formally called an *Euler-Bernoulli beam*[8], three key assumptions and approximations are always made:

- The initial (undeformed) geometry of the beam is used when developing a free-body diagram of forces and calculating the resulting force and moment equilibrium in order to develop the desired governing differential equation. Nowadays in structural engineering this is referred to as a *linear analysis*.

---

[8] A variety of colloquial terms are used as well, with *simple beam* being perhaps the most common in U.S. practice.

14

- Any imaginary vertical plane through the beam cross-section will remain planar and perpendicular to the deformed longitudinal axis of the beam after deflection of the beam. This is the classical 'plane sections remain plane' assumption that is well known to many generations of civil engineers although, as will be seen, planarity is only part of the story.

- Vertical-downward displacements (*deflections*) of the beam, $w(x)$, are relatively small. This is often referred to as *small-deflection theory*.

The resulting differential equation defining the load-deflection behavior of an Euler-Bernoulli beam is the well-known equation:

$$\frac{d^2}{dx^2}\left[EI(x)\frac{d^2w(x)}{dx^2}\right] = q(x).$$

(2.2)

For simplicity in the present discussion and without any loss in the generality of comments to be made, the flexural stiffness of the beam, $EI(x)$, is assumed to be constant (= EI) along the length of the beam which results in the following simplified form of Equation 2.2:

$$EI\frac{d^4w(x)}{dx^4} = q(x).$$

(2.3)

Using the aforementioned matrix-method concept expressed in Equation 2.1, the stiffness matrix, $[S]$, for an Euler-Bernoulli beam is:

$$\begin{bmatrix} \dfrac{12EI}{l^3} & \dfrac{6EI}{l^2} & -\dfrac{12EI}{l^3} & \dfrac{6EI}{l^2} \\ \dfrac{6EI}{l^2} & \dfrac{4EI}{l} & -\dfrac{6EI}{l^2} & \dfrac{2EI}{l} \\ -\dfrac{12EI}{l^3} & -\dfrac{6EI}{l^2} & \dfrac{12EI}{l^3} & -\dfrac{6EI}{l^2} \\ \dfrac{6EI}{l^2} & \dfrac{2EI}{l} & -\dfrac{6EI}{l^2} & \dfrac{4EI}{l} \end{bmatrix}$$

(2.4)

where l = beam length. Not surprisingly, the flexural stiffness of the beam, EI, dominates the stiffness matrix as it is reflected in every element of the matrix.

2.3.2.3 Shear Effects (Timoshenko Beam)

One of the three key assumptions and approximations introduced during formulation of the Euler-Bernoulli beam that has potential ramifications in SSI applications is the plane-sections-remain-plane assumption that is more accurately restated as 'plane sections remain perpendicular to the longitudinal axis'. In reality, internal shear stresses develop within a beam during bending in addition to the normal stresses that produce bending moments. These shear stresses cause planar sections to no longer remain perpendicular to the longitudinal axis of the beam as the beam deflects under transverse load application.

The result of this deviation from perpendicularity is that beam deflections are <u>always</u> greater than those calculated based on the traditional Euler-Bernoulli assumption. The additional component of deflection, i.e. the magnitude of deflection over and above that calculated based on Euler-Bernoulli beam theory, is referred to as *shear deformations* while the primary Euler-Bernoulli component of deflection is called *bending deformations* (Timoshenko and Gere 1972). A beam analysis that considers both bending and shear deformations (deflections) is referred to as a *Timoshenko beam*.

The differential equation defining the load-deflection behavior of a Timoshenko beam for the most basic case of *EI* = constant is:

$$EI\frac{d^4w(x)}{dx^4} = q(x) - \frac{EI}{GA_v}\frac{d^2q(x)}{dx^2}$$ (2.5)

where:
- A_v = *shear area* of the beam[9],
- G = shear modulus of the beam material,

and the remaining terms were defined previously.

Note that some publications express A_v as the product of two separate variables, κ and the actual cross-sectional area, A, i.e. $A_v = \kappa A$. The former variable is referred to as the *Timoshenko shear coefficient, shear coefficient, shear correction factor,* and possibly other terms. This makes it easy to relate shear effects to a single dimensionless parameter that facilitates theoretical discussions and comparisons.

In any event, comparing Equation 2.5 to that for Euler-Bernoulli beam theory (Equation 2.3), it is clear that shear effects are reflected by adding a second term to the right-hand side of the equation. Furthermore, these shear effects clearly relate to the beam loading, $q(x)$, as opposed to the beam deflections, $w(x)$.

The previously noted power and simplicity of the matrix-method alternative can be seen by the fact that the basic equation (2.1) remains unchanged for a Timoshenko beam. Only the stiffness matrix, [*S*], changes by using a new dimensionless parameter, α_v, that incorporates the shear effects and is defined as follows:

$$\alpha_v = \frac{12EI}{GA_v l^2} = \frac{12EI}{G\kappa A l^2}.$$ (2.6)

The stiffness matrix incorporating shear effects can then be expressed as:

$$\begin{bmatrix} \frac{12EI}{(1+\alpha_v)l^3} & \frac{6EI}{(1+\alpha_v)l^2} & -\frac{12EI}{(1+\alpha_v)l^3} & \frac{6EI}{(1+\alpha_v)l^2} \\ \frac{6EI}{(1+\alpha_v)l^2} & \frac{(4+\alpha_v)EI}{(1+\alpha_v)l} & -\frac{6EI}{(1+\alpha_v)l^2} & \frac{(2-\alpha_v)EI}{(1+\alpha_v)l} \\ -\frac{12EI}{(1+\alpha_v)l^3} & -\frac{6EI}{(1+\alpha_v)l^2} & \frac{12EI}{(1+\alpha_v)l^3} & -\frac{6EI}{(1+\alpha_v)l^2} \\ \frac{6EI}{(1+\alpha_v)l^2} & \frac{(2-\alpha_v)EI}{(1+\alpha_v)l} & -\frac{6EI}{(1+\alpha_v)l^2} & \frac{(4+\alpha_v)EI}{(1+\alpha_v)l} \end{bmatrix}.$$ (2.7)

---

[9] Shear area of a beam is neither simply the cross-sectional area of the beam nor an intuitive alternative. Conceptually, shear area is the portion of the actual beam cross-sectional area, A, that contributes to its rigidity in shear and is a function of the cross-sectional geometry of the beam. For a rectangular cross-section as would be appropriate in many SSI applications such as mat foundations, $A_v = \frac{5}{6}A = 0.83A$ (Timoshenko and Gere 1972).

16

Comparing the stiffness matrix for a Timoshenko beam (Equation 2.7) to that for an Euler-Bernoulli beam (Equation 2.4), it is clear that the effect of shear stresses is to <u>reduce</u> the magnitude of most matrix elements (by virtue of the fact that α_v appears only in the denominator of most elements) thus making the beam less stiff (more flexible) in flexure which increases beam deflection under a given load as noted previously.

It is important to understand that shear effects are <u>always</u> present in <u>every</u> beam and thus Euler-Bernoulli beam theory is <u>always</u> approximate on the unconservative side (i.e. <u>always</u> underestimates) with respect to beam deflections. However, both theory and experience indicate that Euler-Bernoulli beam theory produces quite acceptable results for the vast majority of practical applications which is why it has always been and remains a staple of structural engineering education and routine practice. Shear effects generally become significant only as the beam span-to-depth ratio decreases (which is why shear effects are sometimes referred to as *deep-beam behavior*) although the composition and cross-sectional geometry of the beam influence results as well (Roark and Young 1975).

Although the focus here is on beams, it is important to note that there are many SSI applications such mat foundations where the structural member is a plate that, in general, undergoes simultaneous flexure in two orthogonal directions. The same concepts regarding shear effects apply to plates under biaxial flexure. In the case of plates, the terminology used is *Kirchhoff-Love* or *thin-plate* theory (which is analogous to Euler-Bernoulli beam theory in which shear effects are ignored) vs. *Mindlin-Reissner* or *thick-plate* theory (which is analogous to Timoshenko beam theory in which shear effects are considered).

Traditionally, thin-plate theory has been used for SSI analyses, in most cases implied or by default as the issue of shear effects is not even considered or discussed. The relatively few studies to date of the effect of using thin- vs. thick-plate theory for SSI analysis have involved mat foundations. This work suggests that shear effects are not significant, at least for the typical dimensions of mats used for buildings (Horvath 1993c, 1993d; Horvilleur and Patel 1995).

However, this should not be assumed for all mats or SSI applications in general. Recall that shear effects are a function of not only the thickness of the flexural element but its span as well. Thus, there may be applications in which shear effects, which effectively increase beam/plate flexibility and thus increase differential settlements, are important. Given the fact that nowadays it is relatively easy to specify what are referred to as *thick-plate elements* in commercially available structural analysis software (this is illustrated with the case history examples in Chapter 6), it would seem reasonable to simply always use such elements for all SSI analyses involving plates or, as a minimum, explore the sensitivity of results of a given problem to the use of thin- vs. thick-plate elements.

To conclude the discussion of shear effects, for the sake of completeness it is of interest to at least note the other limiting case where bending deformations are insignificant and negligible in magnitude compared to shear deformations. This is the so-called deep-beam behavior noted above. This occurs for a beam that is relatively very deep compared to its span length, *l*. Such structural elements are most often used in building superstructures as what are called *shear walls* or *shear beams* for efficiently stiffening a structure against lateral loading. However, the primary reason for mentioning shear walls/beams here is that shear-only structural elements have found significant use as an imaginary or fictitious element in advanced subgrade models for SSI analyses. It is in this context that shear beams are discussed in Chapter 4. However, as will be seen there is a more analytically efficient way to model shear effects in commercially available structural analysis software so shear beams are typically not used in subgrade models in actual applications although they figure prominently in the theoretical history and development of subgrade models.

2.3.2.4 Nonlinear Effects

2.3.2.4.1 General Behavior

The final structural-mechanics issue discussed involves another of the three key assumptions and approximations inherent in Euler-Bernoulli beam theory, namely that the derivation of Equations 2.2 and 2.3 is based on the initial, undeformed geometry of the beam[10]. The implications of this can be illustrated by a simple example.

Consider the beam shown in Figure 2.3 that is identical to the one shown in Figure 2.2 but with the addition of an axial force, P, that is defined as being positive in compression[11]. Under Euler-Bernoulli beam theory, the axial force has no effect whatsoever on the flexural behavior of the beam and only causes axial-compressive normal stress within, and concomitant axial-compressive normal strains of, the beam[12]. In fact, for an Euler-Bernoulli beam the load P can be increased without theoretical limit (recall that linear-elastic material behavior is assumed) and will never cause buckling of the beam[13].

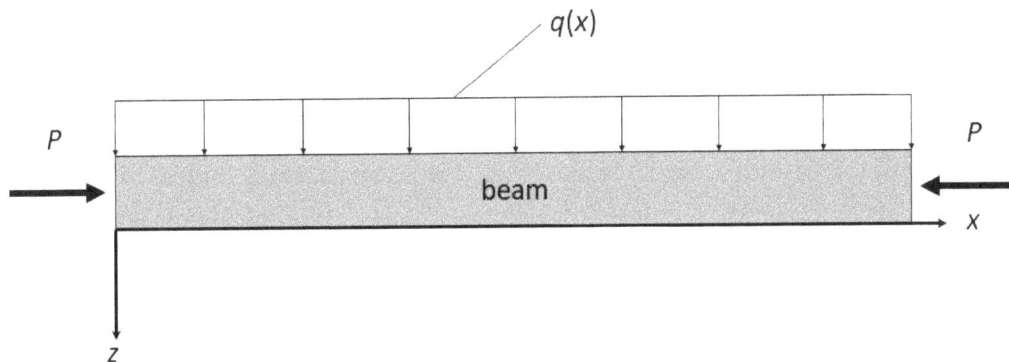

Figure 2.3. Beam with Combined Transverse and Axial Applied Loads.

In reality and independent of the transverse load, $q(x)$, an actual beam (or column) under an axial-compressive load, P, will develop transverse deflections in addition to axial

[10] The third and final fundamental assumption and approximation of Euler-Bernoulli beam theory, small deflections, is not discussed in this monograph. Experience indicates that this is acceptable in SSI applications used to date so there is no need to perform a much more complex large-deflection analysis of flexural elements (beams and plates) although this is routinely possible nowadays with commercially available structural analysis software.

[11] As would be expected, the same basic results are obtained if P is assumed positive in tension. Compression-positive is chosen here as being consistent with the geotechnical sign convention used in this monograph, even for nominally structural topics.

[12] Axial-only effects are sometimes colloquially called 'truss' effects or behavior in structural modeling. This is because a concept analogous to Euler-Bernoulli beam theory (i.e. structural members that have flexural effects only) that is often used in structural modeling is to have structural members (elements) that respond only to axial forces with no flexural effects. This is the assumption made in elementary truss analysis for example. 'Flexural' structural elements and 'truss' structural elements are the basic elements used for linear-analysis modeling in structural engineering.

[13] Note that the same concepts and results apply if the structural element in Figure 2.3 were oriented vertically as a column.

18

deflections, with additional forces and bending moments not forecast by Euler-Bernoulli beam theory. As a result, the beam or column will eventually buckle. This fact has long been recognized by structural engineers.

In the past and particularly in the context of a column subjected to both transverse and axial-compressive loads, this secondary behavior beyond that forecast by Euler-Bernoulli beam theory was called the *P-Δ effect* because the bending moments, additional deflections, etc. were caused by the axial-compressive force, *P*, being displaced transversely (vertically in Figure 2.3) through some distance Δ (parallel to the z-axis in Figure 2.3).

As noted previously, a basic analysis performed using traditional Euler-Bernoulli beam theory is referred to nowadays as a linear analysis so a more-refined analysis performed considering the *P-Δ* effect is referred to as a *nonlinear analysis*. Restated simply, a linear structural analysis is based on the initial, undeformed geometry of a structure and a nonlinear structural analysis takes into account the deformed geometry.

It is of significant importance to this monograph to note that although most commercially available structural analysis software is now capable of performing nonlinear analyses (at least as an optional or add-on capability), this is generally not the default or basic version of such software. Much routine structural analysis and design is still performed using only linear analysis.

As an aside, the way in which nonlinear effects are typically considered in commercial software is by incrementally or iteratively applying linear analysis[14]. There is, apparently, more than one solution algorithm for dealing with nonlinear effects in this manner so nonlinear structural analysis involves a component of art (in terms of the knowledge and experience of the structural analyst) in addition to the basic science built into the software (Nicholas F. Morris, personal communication circa 1990). A discussion of this topic is well beyond the scope of this monograph but it does indicate that the same problem analyzed nonlinearly by different software packages may produce different results.

That having been said, there is one very important and unique exception to the general rule that a nonlinear analysis requires an incremental or iterative linear numerical analysis. Specifically, in one case it is actually possible to obtain a direct, closed-form solution to a nonlinear problem. This is discussed in the following section.

2.3.2.4.2 The 'True' Beam-Column: Basic Formulation and Solution

Many civil engineers, especially geotechnical engineers, call any structural member with both transverse and axial loads as shown in Figure 2.3 a *beam-column*. However, this broad, generic terminology is incorrect. This is because, strictly speaking, the term 'beam-column' only applies to a particular formulation of the problem shown in Figure 2.3.

Specifically, what constitutes what the writer refers to as a 'true' beam-column[15] and makes it unique is not the combined transverse and axial loading per se. Rather, it is that the formulation and derivation of the differential equation defining the behavior of the beam-column is based on the deformed geometry of the beam which has the overall effect of directly and explicitly incorporating the aforementioned *P-Δ* effects into the governing equation.

This is a significant deviation from key assumptions and approximations made for both the Euler-Bernoulli and Timoshenko beam solutions that were based on the undeformed

---

[14] This is identical in concept to the way in which geotechnical finite-element (FE) software solves stress-displacement problems nonlinearly.
[15] The adjective 'true' is dropped but implied for all subsequent uses of the term 'beam-column' in this monograph.

geometry of a beam. As a result, this makes the beam-column inherently a nonlinear problem in the context of the discussion in the preceding section. Furthermore, if certain conditions are assumed[16], it is possible to get the following closed-form solution defining the flexural behavior of a beam-column for the basic case of *EI* = constant and neglecting shear effects:

$$EI\frac{d^4w(x)}{dx^4} + P\frac{d^2w(x)}{dx^2} = q(x).$$ **(2.8)**

Comparing Equation 2.8 to Equation 2.3 for an Euler-Bernoulli beam with similar assumptions concerning *EI*, it can be seen that the flexural effect of the axial force, *P*, on the deformed shape of the beam is reflected in the addition of a second term involving the transverse deflection, *w*, to the left-hand side of Equation 2.8. Also, if *P* = 0 the beam-column equation (2.8) reverts back to the Euler-Bernoulli beam equations (2.3) as expected.

However, the truly intriguing and useful aspect of a beam-column is that it turns what is fundamentally a nonlinear problem into a quasi-linear problem because it is possible to get a relatively simple, exact, closed-form equation (2.8) defining the nonlinear flexural behavior of the beam. Thus, the aforementioned, inherently approximate incremental or iterative numerical-solution techniques normally required for a nonlinear structural analysis are not required in this case.

An even more insightful way to visualize the beam-column problem is using the matrix-method concept. If the shear (Timoshenko-beam) effects that were discussed previously are neglected for simplicity, the stiffness matrix, [*S*], for a beam-column is:

$$\begin{bmatrix} \frac{12EI}{l^3} - \frac{6P}{5l} & \frac{6EI}{l^2} - \frac{P}{10} & -\frac{12EI}{l^3} + \frac{6P}{5l} & \frac{6EI}{l^2} - \frac{P}{10} \\ \frac{6EI}{l^2} - \frac{P}{10} & \frac{4EI}{l} - \frac{2Pl}{15} & -\frac{6EI}{l^2} + \frac{P}{10} & \frac{2EI}{l} + \frac{Pl}{30} \\ -\frac{12EI}{l^3} + \frac{6P}{5l} & -\frac{6EI}{l^2} + \frac{P}{10} & \frac{12EI}{l^3} - \frac{6P}{5l} & -\frac{6EI}{l^2} + \frac{P}{10} \\ \frac{6EI}{l^2} - \frac{P}{10} & \frac{2EI}{l} + \frac{Pl}{30} & -\frac{6EI}{l^2} + \frac{P}{10} & \frac{4EI}{l} - \frac{2Pl}{15} \end{bmatrix}$$ **(2.9)**

and Equation 2.1 can still be used to define the behavior of the beam-column.

Looking at Equation 2.9, note that from a practical perspective the effect of the axial force, *P*, on the flexural behavior of the beam is to modify the flexural stiffness of the beam. A compressive force (positive *P* in Equation 2.9) <u>reduces</u> the magnitude of all elements in the matrix and thus 'softens' the flexural behavior of the beam-column, i.e. makes it more flexible. On the other hand, a tensile force (negative *P* in Equation 2.9) <u>increases</u> the magnitude of all elements in the matrix and thus stiffens the flexural behavior of the beam-column, i.e. makes it more rigid.

Although not reflected in Equations 2.8 and 2.9, shear effects as discussed previously and reflected in the corresponding equations (2.5 and 2.7) for a Timoshenko beam could also be incorporated into the differential equation and stiffness matrix for a beam-column. Note that shear has the same effect on the behavior of a beam-column as it has on a beam with transverse loading only, i.e. it effectively reduces the stiffness of the beam with regard to

---

[16] This involves neglecting axial normal strains and concomitant displacements of the beam due to the applied axial force, *P*. Considering axial strains complicates the problem significantly and requires an iterative numerical solution, similar to nonlinear structural problems in general. Neglecting axial strains turns out to be quite acceptable for most SSI applications.

flexure. Thus, for a beam-column both shear and axial compression act in the same sense to reduce the overall apparent flexural stiffness of the beam-column.

It is of at least academic interest to note that the stiffness matrix for a beam-column provides an alternative perspective for visualizing column buckling. Note that as P increases in magnitude, at some point elements in the matrix (Equation 2.9) become zero which means that the beam loses its flexural stiffness and becomes perfectly flexible. As a result, the transverse displacements, $w(x)$, will be infinite even for an infinitesimally small load transverse load, $q(x)$. In other words, the beam will undergo classical column-type buckling.

For the stiffness matrix shown in Equation 2.9, it can be seen that the critical magnitude, P_{cr}, of the axial force, P, that is required to cause buckling is:

$$P_{cr} = \frac{10EI}{l^2}.$$

(2.10)

Given that the stiffness matrix defined by Equation 2.9 is approximate to some degree because it ignores shear effects and is based on small deflections, this compares very favorably to the theoretical Euler buckling load of:

$$P_{cr} = \frac{\pi^2 EI}{l^2} = \frac{9.9EI}{l^2}$$

(2.11)

for an axially loaded member with ends unrestrained against rotation which corresponds to the problem shown in Figure 2.3 (Timoshenko and Gere 1961).

2.3.2.4.3 The 'True' Beam-Column: Extension to Problems with a Subgrade

There are some interesting issues surrounding use of the beam-column in practice. Although it has historically received treatment in structural-mechanics literature (e.g. Timoshenko and Gere 1961), it appears that the beam-column has seen much more extensive use (in the U.S. at least) not by structural engineers but by geotechnical engineers who have used it since at least the mid-1950s to solve a wide variety of foundation engineering problems that involve SSI. Examples include laterally loaded deep foundations[17], flexible earth-retaining structures (primarily anchored and cantilever sheet-pile bulkheads), mat foundations, and subaqueous conduits (Haliburton 1971, Dawkins 1982, Horvath 1988a). More recently (early 1990s) and with specific and direct relevance to this monograph, the beam-column equation was used as an advanced, multiple-parameter subgrade model proposed by the writer (Horvath 1993e, 1993f). This latter application is discussed in detail in Chapter 4.

Thus, there is an interesting paradox that what is essentially a structural-mechanics problem is much more familiar to, and apparently much more widely used by, geotechnical engineers compared to structural engineers. This appears to be increasingly true as routine structural engineering practice has transitioned to use of the matrix method solved by digital computer. This means that a nonlinear analysis of a beam or column with axial and transverse loading would be performed using an analytical algorithm that is actually an iterative solution of a linear problem formulation as noted previously, thus obviating the need to use the beam-column which is a direct nonlinear formulation.

---

[17] The term 'laterally loaded' is taken in this monograph to include both lateral (horizontal) force and moment loading at the head of a deep-foundation element.

The aforementioned decades-long use of a beam-column in SSI applications as well as its more recent direct use as a subgrade model have required extension of the beam-column problem shown in Figure 2.3 to one with explicit subgrade support as shown in Figure 2.4. The new additions to the problem are an assumed subgrade in contact with the underside of the beam with a concomitant assumed normal contact stress, $p(x)$, between the beam and subgrade. Note that $p(x)$ is not, in general, constant in magnitude along the length of the beam as shown simplistically in Figure 2.4 although in most SSI applications it will be continuous for the entire beam-column length although it does not have to be. As before, the applied transverse load, $q(x)$, is not necessarily constant or continuous.

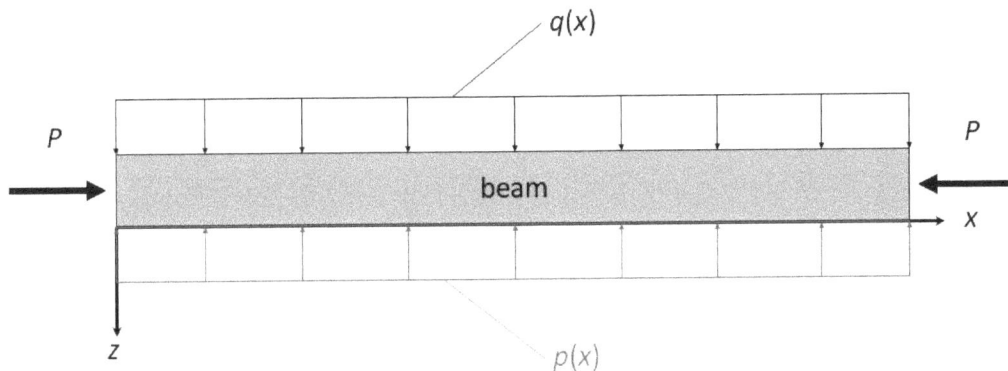

Figure 2.4. Beam-Column with Subgrade Support.

The beam-subgrade contact stress, $p(x)$, shown in Figure 2.4 is called the *subgrade reaction stress* or simply *subgrade reaction* and is analogous to what is typically called the *bearing stress* or *contact stress* in footing applications. Subgrade reaction is a term that appears to be unique to SSI applications (Liao 1991, 1995) and is used throughout the remainder of this monograph.

The differential equation for a beam-column with subgrade support for the basic case of EI = constant is:

$$EI\frac{d^4w(x)}{dx^4} + P\frac{d^2w(x)}{dx^2} + p(x) = q(x).\qquad(2.12)$$

Using the alternative matrix-method formulation:

$$[S]\{d\} + \{p\} = \{q\}\qquad(2.13)$$

where $\{p\}$ is called the *subgrade reaction vector*. Note that the stiffness matrix, $[S]$, is the same as that of a beam-column without subgrade support (Equation 2.9).

As an aside, the special case of $P = 0$ in Equation 2.12 is that of an Euler-Bernoulli beam on a subgrade. In this case, the stiffness matrix used in Equation 2.13 is that of an Euler-Bernoulli beam (Equation 2.4). Note that subgrade support could also be added to a Timoshenko beam or a beam-column with shear (Timoshenko) effects but these alternative problem variants are not pursued in this monograph.

22

2.3.3 Material Issues

2.3.3.1 Introduction

Many, if not most, foundation elements in SSI applications are constructed, at least in part if not entirely, of reinforced Portland-cement concrete (PCC). In SSI applications such as mats where interaction with a superstructure must be considered, reinforced PCC is often used for the superstructure as well.

There are significant material issues that involve the PCC itself as well as the composite reinforced PCC. As will be seen for the case histories discussed in Chapter 6, these issues can significantly impact the flexural stiffness of the various components of SSI applications so need to be discussed in some detail.

2.3.3.2 Reduced Flexural Stiffness: Overview

The inherent flexural stiffness, EI, of a beam is due to two variables:

- the Young's modulus, E, of the material (if homogeneous) or the composite Young's modulus of a section composed of two or more materials; and

- the moment of inertia, I, of the section.

Each of these parameters can vary independently during the service life of an SSI application, primarily due to changes occurring within the PCC component of the foundation element, so it is necessary to consider potential variations during the design phase of a project.

2.3.3.3 Reduced Flexural Stiffness: Reduced Young's Modulus (Creep)

It is widely recognized that PCC exhibits relatively significant creep after it has reached its design strength, primarily as a result of continuing water-content changes within the material. Creep of civil engineering materials under long-term, service-load conditions is more the rule than the exception so this aspect of PCC behavior is not surprising.

In general, a convenient way to both visualize creep as well as deal with it analytically is to consider the Young's modulus, E, of a material undergoing creep on a secant-modulus basis as decreasing as a function of time (Chambers 1984). For PCC, this equivalent modulus reduction is of the order of a factor of two to three and will typically occur over a period of the first few years after construction. More-exact details as to how to calculate this reduction for PCC can be found in Horvilleur and Patel (1995).

While creep-related modulus reduction of PCC is routinely considered by structural engineers for superstructure behavior, it appears to be less often considered for foundation elements in SSI applications such as mat foundations. There is no reason why this should be as it straightforward to deal with (Horvilleur and Patel 1995) and as will be seen for the case histories presented in Chapter 6 the effects can be significant.

Using the matrix-method concept (e.g. Equation 2.4 for an Euler-Bernoulli beam), it can readily be seen that the effect of reducing Young's modulus, E, is a reduction in each of the elements in the matrix. This means that the foundation element, e.g. mat foundation, effectively becomes more flexible with time so for that reason alone differential settlements of the mat and the superstructure it supports will tend to increase over time. Consequently, ignoring creep of the mat PCC and the concomitant reduction in flexural stiffness of the mat is potentially unconservative. This is because controlling differential settlement of the superstructure is often a critical design criterion for a mat-supported structure.

2.3.3.4 Reduced Flexural Stiffness: Cracked-Section Behavior

Another widely recognized material phenomenon concerning PCC is that flexural elements constructed of reinforced PCC will generally crack to some extent within the zone(s) of tensile stress under service loads. This is considered normal behavior. As a result, the structural element cross-section has a reduced effective moment of inertia compared to its original, uncracked moment of inertia (called the *gross moment of inertia*).

This phenomenon of moment-of-inertia reduction due to cracking is referred to as *cracked-section behavior*. For a fully cracked section, a reduction in moment of inertia by a factor of approximately two is common.

As with PCC creep, cracked-section behavior is considered routinely in superstructure analysis but apparently less often for foundation elements in SSI applications such as mat foundations. There is no reason why this should not be done for mats as well[18]. Again, the case history presentations in Chapter 6 illustrate this.

It is again enlightening to use the matrix-method concept (e.g. Equation 2.4 for an Euler-Bernoulli beam) in the present discussion. It can readily be seen that the effect of reducing the moment of inertia, I, is analogous to reducing Young's Modulus, E, as discussed previously for creep as it leads to a direct, proportional reduction in each of the matrix elements. Thus, the primary result of cracked-section behavior is that the structural element will be more flexible that an uncracked one. Consequently, as with creep, ignoring cracked-section behavior of a mat is potentially unconservative for the superstructure as differential settlement is often critical for its design.

In addition, Horvilleur and Patel (1995) found that cracked-section behavior can have a significant influence on calculated bending moments for a mat. Consequently, ignoring cracked-section behavior of a mat is potentially unconservative for the mat as well as it may be structurally underdesigned.

Two aspects of cracked-section behavior make it more complex to deal with analytically compared to creep:

- it cannot be unilaterally applied to all the PCC structural members in a problem as is generally done with creep because it does not impact all PCC members in the same way at the same point in time as does creep and

- it is not an all-or-nothing phenomenon. The transition from gross moment of inertia to fully cracked moment of inertia is progressive as a function of stresses generated within a location-specific cross-section of the structural member by bending moments.

---

[18] Two of the rare documented applications of a cracked-section analysis to a foundation, albeit on a research basis, involved a spread footing (Kerr and Saxena 1977) and a mat (Horvath 1993c, 1993d). Note that the senior author of the Kerr and Saxena paper is the late Prof. William C. 'Bill' Kerr of the Stevens Institute of Technology, not the late Prof. Arnold D. Kerr of New York University and later the University of Delaware. This distinction is made because the paper by W. C. Kerr appears to be a one-off contribution by this author to the subject of subgrade models whereas the contributions to this subject by A. D. Kerr are substantial, substantive, and seminal as discussed at length in Chapter 4.

In most cases, a simplified, conservative analytical approach incorporating cracked-section behavior of the foundation element in SSI applications would simply use the moment of inertia of the fully cracked section which is easily calculated (Horvilleur and Patel 1995). This is what was done for the case histories presented in Chapter 6. However, it is worth noting that a more-sophisticated analysis would be to model the progressive cracking of a reinforced PCC section and the concomitant reduction in moment of inertia from its gross to fully cracked value. There are empirical relationships such as Branson's equation to calculate the effective moment of inertia as a function of developed moment. Such a progressive-cracking analysis would require simulation of incremental loading as well as an iterative solution at each analysis step. Although such an analytical effort is probably excessively exact for most SSI analyses in practice, it has been applied on a research basis to analyses involving mat foundations to at least illustrate the validity of the technique (Horvath 1993c, 1993d).

2.4 STRUCTURAL-ELEMENT WEIGHT

One of the issues that comes up routinely in SSI analyses, especially with mat foundations, is whether or not to include the weight of the foundation element under static loading conditions (dynamic loading is a separate issue that is discussed in the following section). This is not a question that is either easily or unilaterally answered as there are different, sometimes-conflicting issues to consider depending on whether flexure, settlement, or bearing capacity is the focus of the analysis. The nature of the subgrade in terms of its consolidation characteristics also plays a role in the decision-making process. Consequently, this issue requires a surprisingly lengthy and complex discussion.

Appendix A contains a generic, introductory discussion of this issue. Additional, lengthier comments that are specific to mat foundations are presented in Chapter 6 as one of the case history sites used for this monograph involved a subgrade dominated by fine-grain soils where time for various construction steps and soil consolidation due to both stress decrease and increase played significant roles with regard to mat weight.

2.5 DYNAMIC EFFECTS

Up to this point, the discussion in this chapter has focused on applications where the foundation element is subjected to conditions where the applied loading is static, i.e. time-independent, or, based on experience, can be approximately assumed to be so[19]. However, there are SSI applications, typically involving seismic or other types of rapidly applied loading (which may be either transient or steady state), where dynamic-inertia effects of the foundation element must be considered.

Note that, as a minimum, there are two distinct aspects of dynamic loading in SSI applications that need to be considered:

- The behavior of the foundation element itself. Depending on the flexural stiffness of the element, it may be modeled as being either rigid or flexible. In either case, the weight (more technically, the mass) of the foundation element plays a much more substantial and clear-cut role compared to under the aforementioned static loading conditions.

---

[19] Note that this does not imply that the subgrade itself behaves in a time-independent fashion. For example, the subgrade may be subject to classical soil-consolidation effects due to excess porewater dissipation so that even though the loading on the foundation element is static the flexural response of the element is time-dependent.

- The behavior of the subgrade. Note that in this case this involves inertia effects on the subgrade materials and is independent of consolidation effects.

There will also be dynamic effects on any superstructure as well as superstructure-foundation interaction effects but these are separate issues.

This subject, especially for flexible foundation elements, has received considerable attention in the structural engineering and applied mechanics literature. However, experience indicates that the dynamic behavior of foundation elements has had relatively little direct impact on subgrade modeling and subgrade models. Consequently, further consideration of this topic is beyond the needs and thus scope of this monograph. However, the separate issue of time-dependent subgrade response under static loading due to soil-consolidation effects is discussed as this often arises in routine practice.

This page intentionally left blank.

Chapter 3

Genesis of Subgrade Modeling and Models

3.1 INTRODUCTION AND OVERVIEW

This chapter explores how subgrade modeling in SSI applications has evolved over time and, in the process, given rise to the need for subgrade models that persists to the present. The common problem of a mat (raft) foundation is used as an exemplar for this purpose. Although this is a specific type of foundation element and SSI application, for reasons explained subsequently it is arguably the best example of an SSI application to use for this purpose. In addition, the presentation is intentionally broad and generic and thus readily extended to other types of foundation elements and SSI applications, even those with a very different geometry such as laterally loaded deep foundations.

3.2 MAT FOUNDATION ANALYSIS

3.2.1 Problem Components

To illustrate the key concepts of mat foundation behavior, consider the common application where a mat supports a superstructure consisting of some type of building as illustrated conceptually in Figure 3.1. Note that there are three distinct, principal components of the problem:

- superstructure,

- mat, and

- subgrade.

Some comments are required with regard to this figure:

- For simplicity in both illustration and subsequent discussion, the building superstructure is shown without below-ground space so that the top of the mat is flush with the ground surface. This is by no means typical nor does it affect the generality of the subsequent discussion.

- The geotechnical sign convention is used with respect to the coordinate axes but the depth coordinate, z, has its origin at *foundation level* which is defined for the purposes of this monograph as the mat-subgrade interface[20]. This is for mathematical convenience in developing subgrade models for this application, a subject that is addressed in detail in Chapter 4.

---

[20] In geotechnical engineering, the origin of the z axis (i.e. $z = 0$) is usually placed at the ground surface for convenience in calculating overburden stresses and other depth-dependent soil properties and problem parameters.

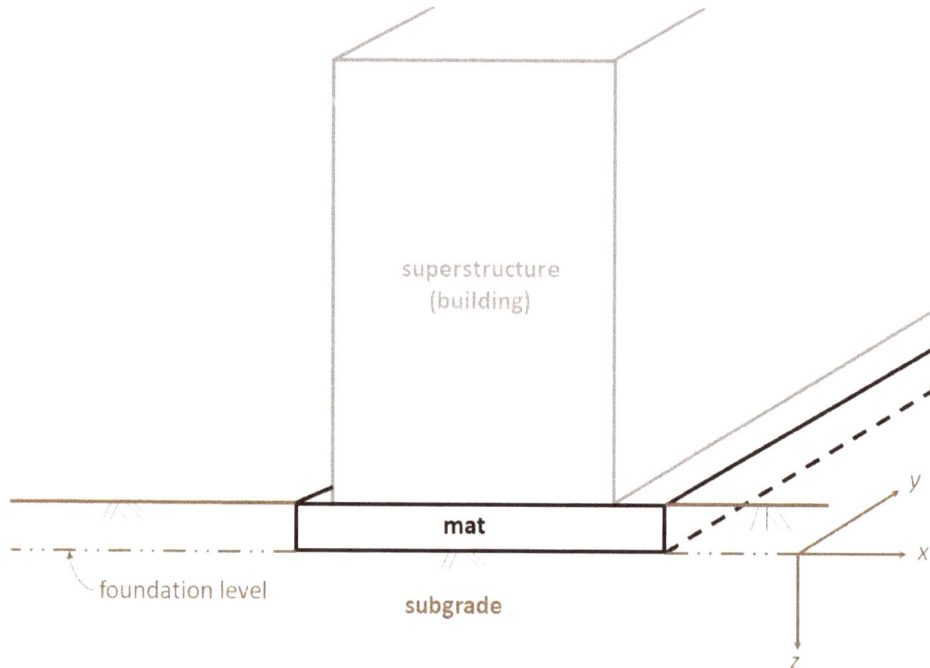

Figure 3.1. Generic Mat Foundation Problem.

- In general, mat flexure can be, and is in general, significant in both horizontal (*x*- and *y*-axis) directions and should always be analyzed for this. However, for simplicity in subsequent presentation and discussion in this chapter as well as Chapter 4, mat flexure in only the *x*-axis direction is illustrated in the various figures (although *x-y* behavior is indicated mathematically).

3.2.2 Key Outcome Parameters of Interest

There are many potential numerical outcomes of interest from an analysis of the problem shown in Figure 3.1. However, for the purposes of this monograph the two parameters of primary interest with regard to the mat are:

- bending moments, which are necessary for structural design, and

- total settlements from which differential settlements that impact design of the superstructure can be calculated.

The reason for identifying and emphasizing these key problem parameters is that history has shown that they have dictated and controlled subgrade modeling and the concomitant development of subgrade models. This will become clear as the following discussion unfolds.

Of equal importance in view of the discussion that follows is noting what is generally unimportant with regard to the mat, specifically, the subgrade reaction across the mat-subgrade interface. The only time when the vertical stress on the subgrade would be of explicit importance is when bearing capacity is an issue which is very rare for mat foundations.

3.2.3 Analytical Strategies for Problem Solution

3.2.3.1 Background

More than any other type of SSI problem in foundation engineering, a mat-supported structure represents an application where SSI is an essential behavioral element and thus should always be considered. The reason is that this is a very complex three-component problem (most other SSI applications have only two components) and the load-displacement behavior of any one of the three components (superstructure, mat, subgrade) in the vertical direction at least is physically linked and interrelated to the behavior of the other two.

Another relevant issue is that due to the complexity of the overall system, there must always be significant involvement by both structural and geotechnical engineers. More than ever, there is simply no one person or civil engineering specialty with the requisite expertise to execute the required state of practice for all three elements of the system. On the other hand, there are many other two-component SSI applications such as laterally loaded deep foundations where one experienced person can adequately address both the geotechnical and structural components of the problem.

In summary, a mat-supported building is arguably the most challenging category of SSI problems in foundation engineering, a sentiment of the writer that is shared by others who have researched and written extensively on the subject:

"The design of...raft foundations undoubtedly represents one of the more difficult technical aspects of civil engineering practice."

- John A. Hemsley, editor of *Design Applications of Raft Foundations*, 2000

It is these technological challenges that makes a mat-supported building the ideal example to use to trace the history of subgrade modeling and, in Chapter 4, the subgrade models that derived from the needs of subgrade modeling.

3.2.3.2 Ideal Solution Alternative

The complex interrelationships between the components of a mat-supported structure means that the ideal solution to the problem shown in Figure 3.1 requires modeling the overall problem as a single entity as shown conceptually in Figure 3.2 (as noted previously, only a 2-D section through the overall 3-D problem is shown for simplicity). Note that the two structural components of the problem, mat and superstructure, are combined into a single structural system (*megastructure*) that bears on the subgrade. As a result of some system of loads applied to the megastructure, there will be a large number of calculated outcomes that includes the vertical downward displacement, $w(x,y)$, (= settlement, ρ) of the megastructure at its foundation level.

It is of both relevance and significance in view of subsequent discussions to note that the subgrade reaction across the mat-subgrade interface is a calculated outcome in this ideal solution alternative. This is noted and highlighted in addition to the previous general observation and comment that the subgrade reaction is generally of secondary importance at best in mat problems and then only if bearing capacity on a fine-grain subgrade requires checking.

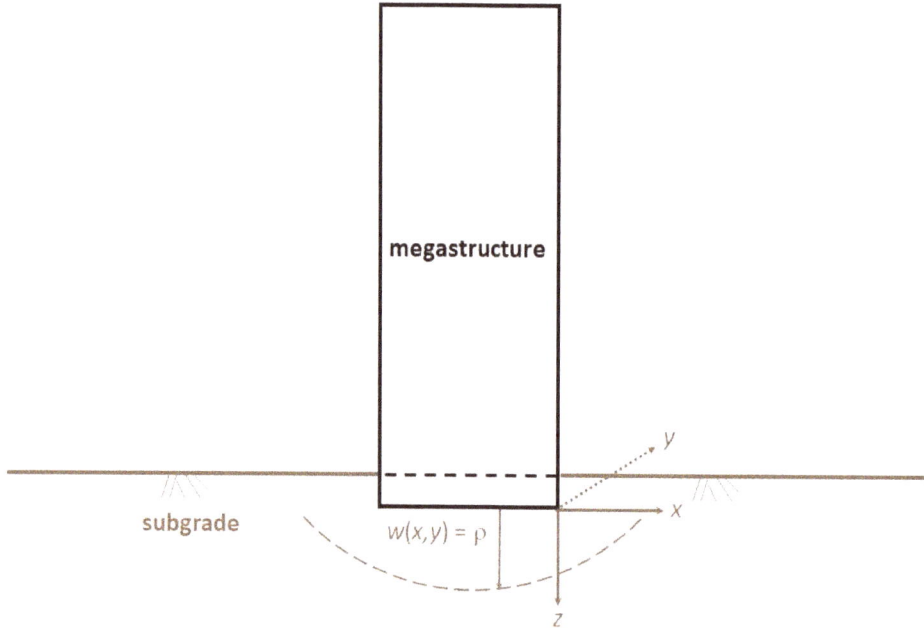

Figure 3.2. Ideal Solution Alternative to Generic Mat Foundation Problem.

Prior to the commercial availability of the digital computer, achieving the ideal solution illustrated conceptually in Figure 3.2 was simply not possible. Since the development of the digital computer, achieving this ideal solution has become achievable in concept and increasingly achievable in reality. However, there have always been and there remain numerous, significant technical issues impeding concept from becoming reality, at least on a routine basis in practice:

- A 3-D numerical analysis is required.

- Both the megastructure and subgrade need to be discretized and solved numerically, most likely using the matrix method and FE method respectively. Note that the FE model for the subgrade component needs to extend horizontally in the x and y directions well beyond the limits of the mat and to a depth z at which settlements of the megastructure can reasonably be assumed to be zero so that the arbitrary boundaries necessary in any FE model (mesh) do not influence the calculated outcomes.

- A single piece of software is required.

- The constitutive models for both the structural and geotechnical components must be state of art and the structural-material capabilities must include both steel and PCC elements.

At the present time, some commercially available software such as *DIANA* from DIANA FEA BV is approaching these capabilities, at least in principle. However, pragmatic constraints still exist:

- The size of the overall numerical model in terms of the number of degrees of freedom may exceed the capability of the software which would require potentially unacceptable compromises to be made in the overall model.

- The computation time for a single analytical run using computers available to the general public would likely be of the order of days, not hours. While access to a supercomputer could reduce the run time, this is not a realistic alternative for most potential users.

- Preparation of the computer input and interpretation of the output would require a team consisting of both geotechnical and structural engineering specialists working with close coordination.

The positive outcome from the significant investment in time and effort that such an ideal analysis would entail is that the forecast results would be expected to be as close to the 'correct answer' as the current state of knowledge can provide. Furthermore, and most relevant to the focus of this monograph, the aforementioned primary results of interest for the mat component (bending moments and total settlements) would be inherently produced as part of the calculated outcomes of the overall system. No special effort, assumptions, or approximations beyond those already used for the overall problem modeling and solution would be required. Finally, for reasons that will become obvious later in this chapter, it is noted that the subgrade reaction across the megastructure-subgrade interface would simply be a calculated, incidental outcome of the overall process.

3.2.3.3 Need for and Role of Subgrade Modeling

While the potential to execute the ideal solution alternative even in routine practice has increased substantially in recent years and will almost certainly only increase further in the future, there are compelling reasons to explore alternative solution approaches:

- At present and for the foreseeable future, it is likely that the majority of mat foundation projects cannot justify the level of effort outlined in the preceding section for the ideal solution alternative. Consequently, a less-demanding solution approach is required.

- Even for world-class projects involving a mat-supported building where the ideal solution alternative might be used for final design, there would undoubtedly be many preliminary-design iterations for which a less demanding solution alternative would be both cost effective and adequate.

- There are many other SSI applications, especially the vast majority that are only two-component problems (foundation element plus the subgrade) for which the ideal solution alternative that treats all problem components and variables in the most analytically rigorous fashion possible is neither justified nor required based on experience using less-demanding solution alternatives. An example would be the common SSI problem of a laterally loaded deep foundation.

- For historical reasons, it is of interest to explore the solution alternatives that were available both prior to the commercial availability of digital computers as well as up until relatively recently when the ideal solution alternative simply was not practical with the computer technology of the time.

32

In summary, there has been and continues to be a significant, meaningful place in both routine practice as well as academic research for SSI problem solution alternatives that are, to varying extents, approximations of the 'reality' that can be achieved with the above-described ideal solution alternative and thus simpler to implement than that ideal alternative.

It is relevant to note that this use of analytical approximations is consistent with the inherent nature of civil engineering which, from its beginnings to the present, has always utilized problem-simplification as its modus operandi. This is because natural systems are, in general, much too complex to model and analyze on a granular level, even with the analytical resources available at present. Note that analytical simplification, if done properly, is elegant in its own right as suggested by a quotation attributed to Albert Einstein:

"Simplicity is the ultimate sophistication."

With this in mind, alternative solutions to the problem shown in Figure 3.1 are discussed in the following sections. First considered is the traditional alternative that had its origins in the pre-computer era and proved to be remarkably resilient in terms of continued use well into the computer era. This is followed by a discussion of what is achievable at the present time within the context of currently available technology.

3.2.3.4 Traditional Solution Alternative

The actual mat foundation problem illustrated in Figure 3.1 is inherently highly complex and statically indeterminate so prior to the availability of digital computers the problem was traditionally broken into its three basic components as shown in Figure 3.3.

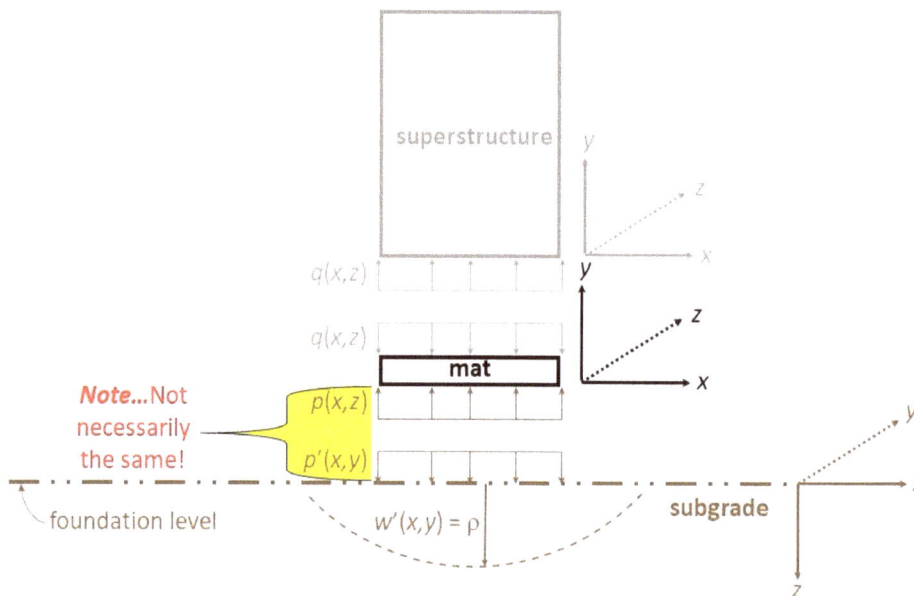

Figure 3.3. Traditional Solution Alternative to Generic Mat Foundation Problem.

The analysis is then conducted in the following sequence of separate steps to effect solution of the overall problem:

- A structural analysis of the superstructure is performed for some system of applied loads (multiple load cases are usually investigated in practice) and, typically, assuming no differential settlement of the column supports. This produces a system of reactions at the base of the superstructure (which is also the top of the mat) that are depicted and referred to qualitatively as $q(x,z)$ in Figure 3.3 (note the use of the structural sign convention for this analytical component). Note that $q(x,z)$ is not necessarily uniform in magnitude or even continuous. In fact, for a building it would typically be a series of individual reactions at the bases of columns.

- A structural analysis of the mat is performed using the applied loads, $q(x,z)$, from the analysis of the superstructure as a known input. Note that this is in and of itself approximate as the superstructure analysis typically ignores differential settlements whereas the mat analysis will, in most cases[21], result in calculated differential settlements. The key aspect of this step is that the subgrade reaction, which is depicted and referred to qualitatively as $p(x,z)$ in Figure 3.3, must be <u>assumed</u> in some manner. The significance of this is elaborated on subsequently. Note that $p(x,z)$ is not necessarily uniform in magnitude although it is generally continuous across the base of the mat. Although not shown for clarity, the mat analysis would produce a pattern of downward displacements of the mat, $v(x,z)$, that would be negative in sign as the mat analysis would typically also be performed using the structural sign convention as indicated in Figure 3.3. As discussed in Chapter 4, these calculated mat displacements were often ignored and the mat analysis used only to forecast the bending moments in the mat for structural design purposes.

- A geotechnical settlement analysis is performed using the geotechnical sign convention to calculate the mat settlements, $\rho = w'(x,y)$, based on some <u>assumed</u> magnitude and distribution of applied loads at foundation level (bottom of the mat), $p'(x,y)$. Logically, the magnitude and distribution of $p'(x,y)$ should be the same as the subgrade reaction, $p(x,z)$, that was assumed for the aforementioned mat analysis. However, this was generally not the case as highlighted in Figure 3.3. This means that the calculated settlements, $w'(x,y)$, are not the same as the calculated mat deflections, $v(x,z)$, as they logically should be. In fact, the two sets of results can differ by an order of magnitude, with mat deflections always be the smaller of the two. The reasons for this are explained in Chapter 4 and illustrated for a specific case history in Chapter 6.

- Geotechnical bearing capacity is a separate geotechnical consideration. Experience indicates that it is not critical in most cases. However, due diligence dictates that it should always be checked, especially if the subgrade consists of fine-grain soil.

For the purposes of the present discussion, the exact analytical methodologies used for each of the above solution components are not important (these details are explored further in Chapter 4) although it is of interest to note that analyses for this alternative are generally 2-D, not 3-D, in nature. This was simply a pragmatic necessity in the pre-computer era. Therefore, the structural analysis of the mat, for example, would generally be done first in one direction (x axis) and then the other (z axis), with the results simply superimposed.

There are some significant implications of note in the traditional solution alternative shown in Figure 3.3:

---

[21] The exception would be if the mat is assumed to be flexurally rigid, an approximation that was sometimes made in the past as discussed further in Chapter 4.

- There is, in general, a lack of compatibility between the vertical displacements of the three components. As noted previously, the superstructure analysis typically assumes no differential settlement of the column bases whereas the mat analysis usually relaxes this requirement as does the settlement analysis of the subgrade. A corollary to this noted previously is that it is not uncommon for the calculated vertical displacements of the mat and subgrade to differ by an order of magnitude. Again, the reasons for this are discussed in Chapters 4 and 6.

- The assumption that the reactions at the base of the superstructure/top of the mat, $q(x,z)$, are the same is inconsistent with the fact that the superstructure is generally assumed not to settle differentially whereas the mat does. As is well known, in statically indeterminate systems such as shown in Figure 3.1 the forces and moments in the structural members comprising the superstructure will be influenced by differential settlements.

3.2.3.5 Root Cause and Defining Characteristic of Subgrade Models

While there are several items of note in the preceding analytical outline, by far the most relevant and important with regard to this monograph concerns the parameter of subgrade reaction, $p(x,z)$, on the base of the mat. Specifically, the pragmatic need to decompose the actual problem (Figure 3.1) into three separate analytical components (Figure 3.3) forces the subgrade reaction to change from being an incidental, calculated outcome of secondary importance in the ideal solution alternative (Figure 3.2) to a primary...indeed _essential_...problem input parameter that must always be known beforehand.

This role reversal concerning the parameter of subgrade reaction is nothing short of a tectonic paradigm shift in SSI applications that has rarely been so noted, emphasized, and appreciated in the past. Furthermore, this role reversal is the root cause of the need for, and the defining characteristic of, subgrade models in all SSI applications that do not employ the aforementioned ideal solution alternative.

In simple terms and as a definition, a _subgrade model_ is an arithmetic expression that defines the resistance between mat and subgrade. As a minimum, a subgrade model is essential for analyzing the mat component of the problem shown in Figure 3.3. Ideally, a subgrade model should also be used for forecasting the settlement of the subgrade component as well as it is logical to have the mat and subgrade settle the same amount in any given problem although this requirement was waived historically as discussed further in Chapters 4 and 6. With reference to Figure 3.3, this means that $p(x,z)$ and $p'(x,y)$ should always be the same.

Although subgrade models play an outsize role not only in the mat foundation example used here but in all SSI applications in which they are used, it is important to note what even a 'good' subgrade model is not. Most importantly, it is _not_ a complete constitutive model for soil. Therefore, subgrade models should never be evaluated using the broad metrics applied to general-purpose soil models but rather only for the limited, focused role for which they are intended to be used.

3.2.3.6 Fundamental Definitions and Basic Parameters Related to Subgrade Reaction

Returning to the parameter of subgrade reaction, because $p(x,z)$ plays such a crucial role in the analysis of the mat component shown in Figure 3.3, it turns out to be useful to define a new problem parameter, $k(x,z)$, that reflects the subgrade stiffness and is herein defined as the *coefficient of subgrade reaction*[22]. It is defined as follows:

$$k(x,z) = \frac{p(x,z)}{v(x,z)}.$$

(3.1)

It cannot be emphasized too strongly (the reasons will become clear in Chapter 4) that k (as it will be referred to for simplicity) is completely general and generic in its fundamental concept and basic definition. It is simply the ratio of the subgrade reaction to settlement at any given point across the bottom of the mat. That having been said, the <u>interpretation</u> and concomitant importance of k varies widely depending on the specific application. For example, when used in the context of the traditional solution alternative shown in Figure 3.3, k is highly subjective in its evaluation in a particular application. This is because it involves the problem parameter of subgrade reaction, p, that is an input parameter in this application and thus must be defined a priori. On the other hand, in the generic problem shown in Figure 3.1 and its ideal solution alternative shown in Figures 3.2, the nature of k is a completely different. It is a calculated result of arguably secondary interest and importance because it is the ratio of two calculated results.

This, then, is the root source and cause of confusion surrounding the coefficient of subgrade reaction, k, in SSI applications. While it always has the same fundamental definition, it has a wide range of interpretations, implications, and importance depending on the specifics of an application. This will become clear in Chapter 4.

Additional comments are warranted concerning the generic coefficient of subgrade reaction, k, as it has considerable consequence for the discussion of subgrade models in Chapter 4. It should be obvious from the actual problem (Figure 3.1) that both the magnitude and distribution of both the subgrade reaction, p, between the mat and subgrade and settlement of the mat, v, are influenced by the flexural stiffness, D, of the mat. This means that both the magnitude and distribution of k are functions of D as well. This can be illustrated using the simplistic 2-D problem shown in Figure 3.4.

This figure shows the two limiting cases of mat flexural stiffness, $D = 0$ (perfectly flexible) and $D = \infty$ (perfectly rigid). The intermediate values of $0 < D < \infty$ that exist with an actual mat foundation produce transitional results between these limiting cases. The applied load, q, from the superstructure is assumed to be a uniformly distributed load for simplicity and the same in each case. The subgrade reactions, p, and settlements, v, are what would be expected for idealized linear-elastic subgrade.

---

[22] The term 'coefficient', not 'modulus', is preferred as the fundamental dimensions of $k(x,z)$ are always force per length cubed. Use of the term modulus, which is frequently done with this parameter (see, for example, Hetényi (1946) who used the notation k_0 and called the parameter the *modulus of the foundation*), should be discontinued as the term modulus is generally reserved for stiffness parameters with dimensions of force per length squared, e.g. Young's modulus, shear modulus, etc. Another compelling reason why this distinction in terminology is neither trivial nor academic is that there is a distinctly different SSI parameter called the *modulus of subgrade reaction*, with dimensions of force per length squared, that was defined and used by the late Prof. Aleksandar Vesic who conducted significant research into subgrade modeling and models involving beams. In Vesic's case, the modulus of subgrade reaction, K, = kb where b = the width of the beam. Unfortunately, this notational difference for modulus of subgrade reaction (K) as opposed to coefficient of subgrade reaction (k) is not universal. For example, Hetényi (1946) defines modulus of subgrade reaction in the very first pages of his scholarly monograph but with the notation $k = k_0 b$. The combination of terminological and notational inconsistency from one author to another is a significant part of the reason why the overall subject of subgrade modeling and models is so confusing to most users.

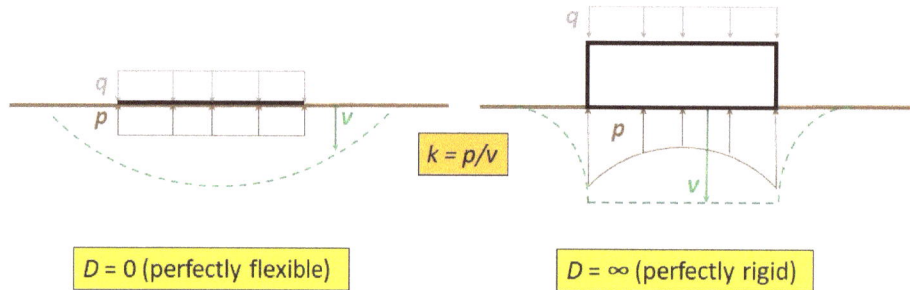

$k = p/v$

$D = 0$ (perfectly flexible) $D = \infty$ (perfectly rigid)

Figure 3.4. Qualitative Effect of Mat Stiffness on Subgrade Reaction and Settlement.

It is visually apparent that for the same applied load, q, that both p and v are very dependent on D which implies that k is dependent on D as well by virtue of the relationship shown in Equation 3.1 and Figure 3.4. It is also visually obvious that in both limiting cases there are settlements, v, even in areas where $p = 0$ and thus $k = 0$.

The dependency of the overall SSI problem in general and k in particular on the flexural stiffness, D or EI, of the foundation element was recognized by researchers at least by the middle of the 20th century. However, this fundamental knowledge does not appear to have been disseminated to practitioners to the extent that it should have after more than half a century.

This same research found that it was useful...particularly so in the pre-computer era in which this research was done when closed-form arithmetic solutions were the only possible outcomes and thus defined the state of art for use in routine practice...to define a new parameter that reflected the relative stiffness between the foundation element and subgrade, something that k alone does not and can never do. Unfortunately, this research suffers from assumptions necessary to achieve a closed-form solution, primarily that k is constant in a given problem which the above simple discussion centered on Figure 3.4 shows is clearly never the case. However, there is still historical and intellectual value to discuss this new parameter.

The first body of research was by Hetényi (1946) who was primarily interested in the behavior of beams. He defined a parameter, λ, that was called the *characteristic of the system* or simply *characteristic* and has dimensions of length^{-1}:

$$\lambda = \sqrt[4]{\frac{3k}{Et^3}}$$

(3.2)

where E and t are the Young's modulus and thickness of the beam respectively. The limitation of Equation 3.2 is that it assumes a beam with a rectangular cross-section in addition the aforementioned assumption of k = constant.

Although λ appears extensively in the closed-form solutions developed by Hetényi, of more-direct practical interest is its reciprocal, $1/\lambda$, that Hetényi called the *characteristic length* as it has dimensions of length. The specific pragmatic value of the characteristic length is that it can be used as the basis for quantifying behaviors of potential practical use. For example, for an isolated point load applied to a beam, the horizontal distance along the beam from the point of load application to a point at which the beam settlement from that load is approximately zero is given theoretically by:

$$\left(\frac{3\pi}{4}\right)\left(\frac{1}{\lambda}\right) \cong 2.4\left(\frac{1}{\lambda}\right). \tag{3.3}$$

Terzaghi (1955) broadened these concepts to include plates by referencing work published in 1926 by the late Prof. Harold M. Westergaard (Westergaard 1926) that was related to 'rigid' (PCC) pavements. Again, the Westergaard/Terzaghi results are constrained by the necessity to assume that k = constant.

With plates, the conceptual equivalent of the characteristic length, $1/\lambda$, is called the *radius of stiffness*, r_0:

$$r_0 = \sqrt[4]{\frac{D}{k}} = \sqrt[4]{\frac{Et^3}{12(1-\nu^2)k}} \tag{3.4}$$

where D is the flexural stiffness of a plate of thickness t and ν is the Poisson ratio of the plate material that has Young's modulus E.

Assuming that the plate is composed of PCC with $\nu = 0.15$, Equation 3.4 simplifies to:

$$r_0 \cong \sqrt[4]{\frac{Et^3}{12k}}. \tag{3.5}$$

As with beams, this parameter can be used to draw several practical inferences. For example, for a plate (e.g. mat, slab-on-grade, PCC pavement) subjected to an isolated point load, the radial distance from the point of application of the load to a point at which settlements of the plate due to that load are effectively zero is called the *range of influence*, R, and is given theoretically by:

$$R = 2.5r_0 \cong \sqrt[4]{\frac{10Et^3}{3(1-\nu^2)k}} \cong \sqrt[4]{\frac{10Et^3}{3k}} \tag{3.6}$$

assuming that the plate is composed of PCC.

These relationships involving the $1/\lambda$ and r_0 parameters were used extensively in older literature in the pre-computer era when only closed-form solutions were available for use and every effort was made to extract useful information from those solutions. For example, mat foundations for two of the case histories presented and discussed in Chapter 6 were originally designed in the 1960s timeframe using published solutions based on these parameters. The original reference for these case histories (DeSimone and Gould 1972) contains an excellent, detailed discussion of how an experienced foundation designer of that era artfully combined these solutions and parameters with engineering judgment in order to create a final design that reflected the then state of art.

The need to use these parameters has all but disappeared in the current computer era. However, as will be seen later in this chapter they have been replaced with alternative non-dimensional parameters that seek to quantify the relative stiffness between a mat and subgrade. Nevertheless, for all their simplicity (due to the k = constant assumption on which they are based) the $1/\lambda$ and r_0 parameters still provide informative insight and reinforce the very important fact that SSI behavior is a function of relative foundation-subgrade stiffness of which k is but one component.

3.2.3.7 Modern Solution Alternatives

3.2.3.7.1 Background and Overview

Even with the current wide availability of digital computers and software in routine civil engineering practice, the traditional, pre-computer era solution alternative of decomposing a mat foundation problem into three separate analytical components as shown in Figure 3.3 and discussed in the preceding sections persists to the present. This is totally unnecessary and not recommended as more-accurate alternatives exist within the capabilities of routine practice. Therefore, the traditional solution alternative is rejected as a viable alternative for continued use in modern practice, a sentiment that is shared by others who have worked extensively with SSI applications, especially mat foundations:

> *"Until fairly recently, there was little alternative but to proceed on the basis of greatly simplifying assumptions combined with rudimentary analysis. But although many such designs were developed with remarkable success, the limitations of this traditional approach cannot be disregarded and often are unacceptable in modern practice."*

> - John A. Hemsley, editor of *Design Applications of Raft Foundations* (2000)

In the writer's opinion, striking a balance between the theoretical rigor applied to a problem and the accuracy and precision of the forecast outcomes should always be guided by another quotation that is attributed to Albert Einstein on the subject of analytical simplification:

> *"Make things as simple as possible but no simpler."*

With particular reference to SSI applications, this means that problem simplification and concomitant solution alternatives should not be temporally fixed but continuously adjusted over time as technology changes and evolves.

With this in mind, three alternative solution approaches more modern than the traditional approach discussed previously have been defined by the writer. They fall into three broad categories:

- structural,

- geotechnical, and

- pseudo-ideal.

3.2.3.7.2 Structural Alternative

The structural alternative analytically combines the superstructure and mat into one structural model (megastructure) as shown in Figure 3.5. This duplicates one key element of the ideal solution alternative shown in Figure 3.2 in that it allows for extremely accurate, state-of-art 3-D modeling and concomitant analysis of all structural components with commercially available structural analysis software which is why the structural sign convention is shown in Figure 3.5.

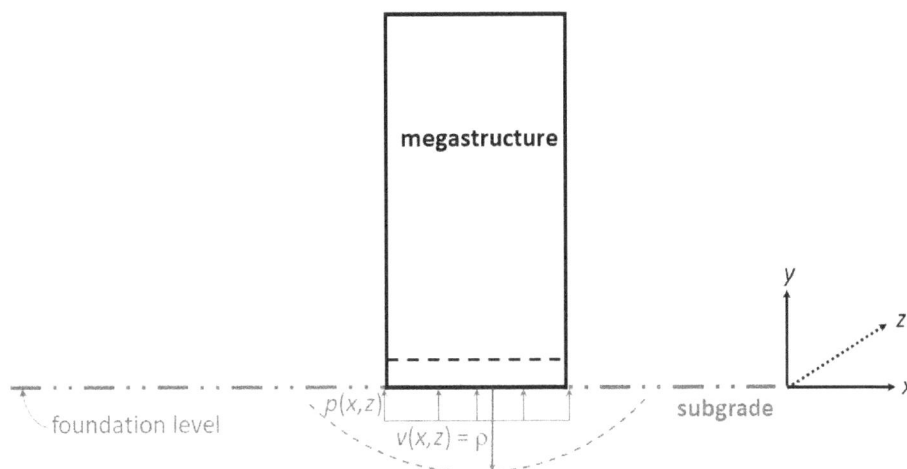

Figure 3.5. Structural Solution Alternative to Generic Mat Foundation Problem.

This alternative also rigorously and correctly accounts for superstructure-mat interaction effects. Numerous studies (e.g. Burland et al. 1977, The Institution of Structural Engineers 1989, Ulrich 1991) have found this interaction to be important to include as a rule. Ignoring this interaction is one of many limitations and resulting weaknesses of the traditional solution alternative shown in Figure 3.3.

The primary limitation and potential shortcoming of the structural alternative is that the subgrade reaction on the base of the megastructure (i.e. bottom of the mat), $p(x,z)$, must be defined beforehand using a subgrade model that can be accommodated by the commercially available structural analysis software used for the megastructure. This is a classic example of what was noted in Chapter 1 that in some SSI applications such as mat foundations that the technical requirements of the structural engineering aspects can completely dominate and 'drive' the overall problem-solution methodology. Furthermore, the subgrade reaction must be defined in a manner that will produce a forecast of the actual megastructure settlements, $v(x,z)$, not pseudo-settlements of no practical value as with the traditional alternative.

3.2.3.7.3 Geotechnical Alternative

The overall objective of the geotechnical alternative is to rigorously model both the mat and subgrade in one analysis as shown in Figure 3.6. This duplicates one key element of the ideal solution alternative shown in Figure 3.2 in that it allows for state-of-art modeling and analysis of the actual geotechnical continuum. The analytical sophistication of the mat component depends on the specific geotechnical software used.

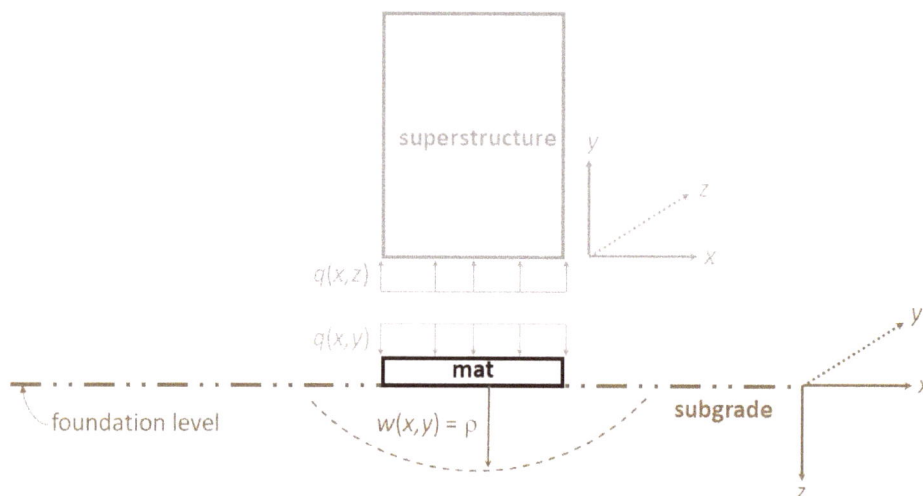

Figure 3.6. Geotechnical Solution Alternative to Generic Mat Foundation Problem.

Nowadays, the geotechnical analytical state of art is defined by 3-D FE analyses using commercially available geotechnical engineering software such as *DIANA* and *PLAXIS 3D*. The use such programs in routine practice is increasingly possible although execution times for just a single analytical run can be lengthy (i.e. days) as was the case with the ideal solution alternative discussed previously. Consequently, computer-based but less-resource-demanding analytical methodologies that defined the geotechnical state of art in the relatively recent past are still potentially useful. Examples include software based on closed-form solutions for a layered, linear-elastic system (one such example is discussed and used in Chapter 6) and chart solutions developed using the boundary-element method (BEM) such as those in Fraser and Wardle (1976).

Note that a separate analysis of the superstructure is required, presumably using commercially available structural analysis software. This allows for an accurate, state-of-art 3-D analysis of the superstructure but with the significant omission of mat-superstructure interaction. This is because the reaction loads, $q(x,z)$, from the superstructure that would be applied to the top of the mat as $q(x,y)$ in the geotechnical analysis would presumably neglect differential settlement. As noted above, mat-superstructure interaction is very important to consider in most cases if the correct mat settlements and bending moments as well as correct forces and moments within the superstructure are to be calculated. Thus, the geotechnical alternative has the obvious shortcoming of the lack of ability to model superstructure interaction effects directly[23]. However, the biggest overall shortcoming of the geotechnical alternative is that two separate, relatively sophisticated, time-consuming analyses muse be performed for the subgrade + mat and superstructure yet the outcomes of this effort are short-changed by the compromises made with each in terms of not linking settlements between the two analyses.

For historical purposes, it is of interest to note that one of the outgrowths of early research into the geotechnical solution alternative was the development of empirical

---

[23] There have long been approximate methods to account for mat-superstructure interaction, some of which have been incorporated into recommended practices published by industry and professional organizations such as the American Concrete Institute (ACI). These methods typically involve increasing the flexural stiffness of the mat using some empirical relationship that depends on the estimated flexural stiffness of each floor within the superstructure.

relationships for what are generically called *relative stiffness factors* and usually given the parameter notation K. This dimensionless parameter is defined and used to estimate the flexural rigidity/flexibility of a mat relative to its subgrade, primarily as it affects differential settlements but, in some cases, bending moments within the mat as well. As such, this parameter is an updated version of the parameters of characteristic length, $1/\lambda$, radius of stiffness, r_0, and range of influence, R, discussed previously. The difference is that the Young's modulus of the subgrade is used in the evaluation of K as opposed to the coefficient of subgrade reaction that is used for these other, earlier parameters.

Different definitions of K have been proposed over the years by Brown (1969), Fraser and Wardle (1976), Horikoshi and Randolph (1997), and others. However, they all have elements in common:

- The flexural stiffness of the mat as defined by the Young's modulus of the mat material and the mat thickness cubed which, for all practical purposes, is the classical parameter of the flexural stiffness of a plate, D. A subtle point is that this implies a gross (uncracked) section of the mat. The significance of this is explored in Chapter 6 for three case histories.

- The stiffness of the subgrade as defined by its Young's modulus.

- The larger the value of K, the stiffer the flexural behavior of the mat. Limiting values of K are zero (perfectly flexible) and open-ended for increasing rigidity.

Unfortunately, each definition of K is unique to a given researcher as the empirical equations used to define K differ (other parameters such as the Poisson ratio of the mat and subgrade materials and the width and length dimensions of the mat may or may not be included). Thus, a given equation for K is only of practical use with the published results that accompany it.

3.2.3.7.4 Pseudo-Ideal Alternative

The pseudo-ideal alternative is shown in Figure 3.7 and incorporates elements of both the structural and geotechnical alternatives. Specifically, an analysis of the mat + superstructure (megastructure) is performed using commercially available structural analysis software. However, no subgrade model is explicitly incorporated into the structural model. Rather, a separate, parallel geotechnical analysis is performed using commercially available geotechnical analysis software. The two analyses are coordinated in an interactive, trial-and-error fashion by matching the subgrade reactions ($p(x,z)$ and $p(x,y)$ respectively) as well as the settlements ($v(x,z)$ and $w(x,y)$ respectively) for the two analyses until the calculated outcomes of each analysis converge to the same values within some precision deemed acceptable by the design teams performing the analyses.

In practice, this matching is generally achieved by the structural engineer placing independent axial springs at several arbitrary locations across the bottom of the megastructure model. The locations of these springs correspond to load points on the surface of the geotechnical model. The stiffnesses of these springs in the structural model and the magnitudes of the corresponding loads in the geotechnical model are then adjusted in a trial-and-error manner until the calculated force and settlement results match as desired.

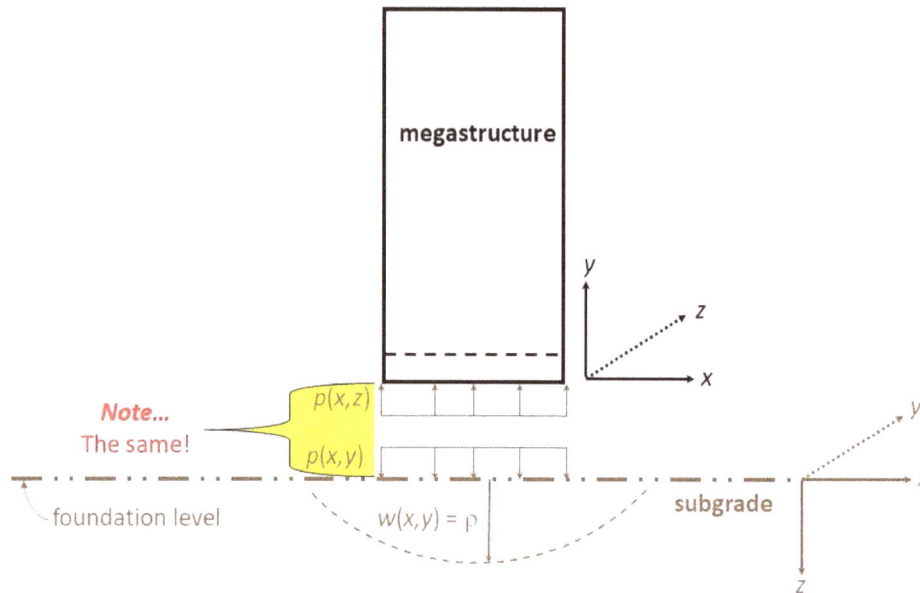

Figure 3.7. Pseudo-Ideal Solution Alternative to Generic Mat Foundation Problem.

3.2.3.7.5 Assessment of Alternatives

In assessing and ranking the three modern solution alternatives outlined above for their potential use in routine practice, a logical initial reaction might be that the last (pseudo-ideal) would be the one of choice as it offers the potential for modeling both the structural and geotechnical components of the problem as accurately as commercially available software technology allows. With specific regard to the issue of subgrade modeling that is a focus of this monograph, this alternative allows explicit modeling of a project-specific geotechnical continuum with the minimum of approximations and compromises while at the same time allowing accurate analysis of the megastructure, including settlement-related mat-superstructure interaction effects.

Supporting this position is the fact that the pseudo-ideal alternative actually has a surprisingly long history of use that is approaching a half-century in length, apparently beginning as what is called the *Discrete Area Method* in the published literature (Ulrich 1991). It was reportedly (Banavalkar and Ulrich 1984) used as early as circa 1970 (Focht et al. 1978) and in the first two decades of its use reportedly produced good results on a number of major mat foundation projects involving high-rise buildings in the U.S. (Ulrich and Jacob 1979, Banavalkar and Ulrich 1982, Ulrich 1991). This is all the more notable as the geotechnical software used on these projects is believed to have been far more primitive analytically compared to the 3-D FE software that is currently available[24].

---

[24] The software used by Ulrich and colleagues for the geotechnical analyses is not known to the writer with certainty. It is believed to have been capable of solving the problem of a 3-D, multi-layered, linear-elastic continuum, with arbitrary zones or areas of different uniform vertical stresses (= subgrade reactions) applied to its surface in a checkerboard fashion. Such software is known to have been among the first developed for geotechnical applications and operating on mainframe digital computers in the 1960s. One pioneering example of such software was the *ICES SEPOL* (**I**ntegrated **C**ivil **E**ngineering **S**ystem **Se**ttlement **P**roblem **O**riented Computer **L**anguage) package developed at the Massachusetts Institute of Technology (Schiffman et al. 1970).

On the negative side, the reported history of use of the Discrete Area Method apparently involved a very limited number of U.S. geotechnical and structural engineering consulting firms working together on projects. Furthermore, an Internet search performed during preparation of this monograph surprisingly revealed no published mention of this method (other than by reference in some of the writer's earlier publications) since the early 1990s even though one of the principal geotechnical team members (Edward J. Ulrich Jr.) is apparently still in professional practice. This suggests that there were pragmatic issues of some unknown nature that inhibited its continued use in the longer term.

Another consideration is that while the pseudo-ideal alternative may have a workable methodology for mat foundations as the Discrete Area Method, it is not conducive to use as a general methodology for other SSI applications. Consequently, the pseudo-ideal alternative is ruled out as a practical alternative for routine use in practice and thus further consideration in this monograph. However, it remains, in concept at least, a viable alternative for those willing to work within its constraints. The writer is aware of anecdotal information that the generic concept of the pseudo-ideal alternative has been employed by other teams of geotechnical and structural engineering design professionals up to the present, just not under the specific banner of the Discrete Area Method.

Of the two remaining alternatives (structural and geotechnical), it is the writer's opinion that only the first (structural) approach of working with and within the modeling constraints of commercially available structural analysis software represents a reasonable and practical approach to the problem based on the current state of computational ability available in routine practice. There is no doubt based on the published literature (e.g. Fraser and Wardle 1976, Tomlinson 1986) that the geotechnical alternative was very attractive to practitioners in the past when computer technology was much less advanced and available than in the present. However, it would appear that the temporal window in which the geotechnical alternative was viable has closed for good. Furthermore, the structural alternative is sufficiently general that it can be applied to SSI applications other than mat foundations in a straightforward and efficient way. Consequently, it is the only alternative considered in detail in the remainder of this monograph.

A final comment that applies to all the solution alternatives discussed in this chapter is the issue of boundary conditions for the subgrade component. Note that these are in addition to the usual boundary conditions for the foundation element (e.g. mat) that are well-known and not discussed further in this monograph.

In all of the preceding figures, it is clear that, in general, the subgrade reaction has subgrade settlement effects beyond the horizontal limits of the mat. While this is reflected pictorially by settlements extending beyond the limits of the mat, what is unseen but nevertheless present is effects within the subgrade itself. Thus, in the discussions that follow in Chapter 4 concerning specific subgrade models, the issue of boundary conditions must always be addressed explicitly.

3.3 SUMMARY

To summarize and close out this chapter as well as to provide a logical segue to the following chapter, the position taken in this monograph is that, given the current state of analytical technology available to the average design professional in practice, a modern solution alternative remains the overall optimum choice for SSI problems in general. The ideal solution alternative of analyzing an entire SSI application exactly in one numerical model, while increasingly achievable, has numerous pragmatic limitations that limit its use to a relatively small fraction of SSI applications and projects. At the other end of the analytical spectrum, the traditional solution alternative that harks back to the pre-computer era has not been intellectually or analytically defensible for decades.

Of the three modern solution alternatives that were presented and discussed, the structural alternative is clearly the alternative of choice in most cases. This is true not only for the mat foundation problem that was used to illustrate the various solution alternatives but for other SSI applications such as the common laterally loaded deep-foundation problem as well.

The key issues to be addressed in the remainder of this monograph are the subgrade models that are used for the structural solution alternative. Specifically, the challenge in SSI applications in general is how to strike a balance between theoretical accuracy and ease of use to model the subgrade reaction, p, that is required input for the structural alternative shown in Figure 3.5.

In the specific example of a mat-supported building as used in this chapter for illustrative purposes, the subgrade model must be achievable within the constraints of commercially available structural analysis software. In other SSI applications such as laterally loaded deep foundations where there is generally purpose-written software, there are other characteristics and constraints placed on the subgrade model.

How to rationally develop and assess subgrade models is the overall topic pursued in Chapter 4. However, before proceeding to this chapter there is an additional topic to address in an introductory manner as it has significant relevance to the selection of subgrade models in the current state of practice that is presented and illustrated with case history examples in the remainder (Chapters 5 and 6) of this monograph.

3.4 SITE CHARACTERIZATION

Although this monograph focuses on the development and mechanics of subgrade models, it is important to note that no geotechnical model, regardless of its degree of theoretical sophistication, is of any practical value if the input parameters required for the model are either difficult/impossible to quantify with reasonable cost and effort or quantitatively significantly in error. With particular regard to the latter, it is well-known and long established that even the most sophisticated soil or subgrade model cannot correct for inadequate/incorrect subgrade properties. Therefore, for the sake of completeness it is important to emphasize the need for an appropriate, site-specific geotechnical site characterization study for every project where subgrade modeling and an SSI analysis is to be performed.

This issue is raised and the last point is emphasized as historically it was long thought that the simple subgrade models that are still in widespread use to the present did not require site-specific geotechnical investigation beyond a gross distinction of 'sand vs. clay' for the subgrade soil and perhaps a subjective assessment of its consistency, i.e. 'loose vs. dense' or 'soft vs. stiff' as appropriate. As will be illustrated in Chapters 5 and 6, this simplistic approach to site characterization is never sufficient.

As discussed in detail in Chapters 5 and 6, the most important aspects of site characterization for the SSI analysis of mat foundations and most other foundation elements are overall site stratigraphy and the variation of soil stiffness (modulus) with depth. Although Young's modulus, E, (or its special case of the constrained modulus, D[25]) was used historically, in recent years there has been a trend toward using shear modulus, G, as the latter is relatively

---

[25] Unfortunately, this geotechnical parameter has the same notation as the structural parameter for the flexural stiffness of an elastic plate. Both parameters are used in this monograph but it should be obvious from the context of usage which parameter is intended.

easily determined using modern in-situ testing devices such as the seismic piezocone (sCPTu). Stress-state parameters, especially *yield stress*[26], are also important as the operational soil stiffness is a function of whether the soil is loaded below or beyond the yield stress among other things.

---

[26] This parameter was formerly referred to by various names such as *preconsolidation pressure/stress*, *natural prestress*, and *maximum past vertical effective pressure/stress*.

This page intentionally left blank.

Chapter 4

History of Subgrade Models

4.1 INTRODUCTION AND OVERVIEW

This chapter focuses on the evolution and development of subgrade models that can be used to define the subgrade reaction, $p(x,z)$, in the structural solution alternative for SSI problems that was illustrated in Figure 3.5 using the classical exemplar problem of a mat foundation. In some cases, the alternative matrix notation, $\{p\}$, that was introduced in Chapter 2 will be used when illustrating how a subgrade model can be incorporated into structural analysis software that nowadays is almost always based on the matrix method of problem formulation and solution.

The one change made in this chapter is that the geotechnical sign convention is used throughout. This is to be consistent with both the historical perception and development of subgrade models as being special-purpose, limited-application soil models as well as earlier published works by the writer. As a result, the subgrade reaction in the aforementioned structural solution alternative becomes $p(x,y)$...hereinafter denoted simply as p for simplicity...for the mat foundation application and other SSI applications involving a foundation element with a horizontal orientation such as a slab-on-grade. Similarly, settlements become $w(x,y)$ or w for simplicity. SSI applications with other geometries, such as laterally loaded deep foundations, may use other notation.

4.2 FUNDAMENTAL CHARACTERISTICS OF SUBGRADE MODELS

Subgrade models as defined in this monograph possess several necessary characteristics:

- They are not 'true' (i.e. complete) soil models although as will be seen in this chapter, a goal of more-advanced subgrade models is to at least, in the limit, trend to and approach the behavior of what is arguably the simplest true soil model, that of a linear-elastic solid. Rather, subgrade models are intended to be tradeoffs between theoretical completeness and ease of use for very specific applications to provide estimates of the subgrade reaction, p, that will produce acceptable results of key parameters in that targeted application. For example, as discussed in Chapter 3, the key parameters of interest for a mat foundation are bending moments in and total settlements of the mat[27]. Furthermore, because subgrade-model parameters are sometimes calibrated for a very narrow, well-defined application, the same subgrade model can have different parameter calibrations for other applications.

- They analytically reduce the length dimensions of the subgrade under consideration, generally by one variable but in some applications by two variables. This is the simplifying feature that makes subgrade models so appealing to use in foundation

---

[27] As discussed later in this chapter, in the earliest use of subgrade models with mats only the former parameter (bending moments) was sought.

48

engineering. For the general mat foundation problem illustrated in Figure 3.1, the depth variable (z) is eliminated. It is assumed that some function of x and y over the surface of foundation level ($z = 0$) can be postulated that will adequately account for the missing depth effects. Applications where two variables are eliminated include mat foundations where the overall megastructure is relatively rigid in one dimension (e.g. y) so that mat flexure in one direction (e.g. x) need only be considered, as well laterally loaded deep foundations where the historical development of subgrade models has resulted in the need to only consider one horizontal dimension (y) for the subgrade in the problem formulation and solution.

- They produce estimates of subgrade reactions and displacements in only one axis direction as the result of applied loads acting only in the same axis direction. This direction is always that in which the dimension (if one) or a dimension (if more than one) was eliminated. For example, in the general mat foundation problem illustrated in Figure 3.1, the subgrade reaction, p, is the mat-subgrade contact stress in the z-axis direction acting upward on the bottom of the mat and the settlement, w ($= \rho$), is also that in the z-axis direction.

4.3 UNIFYING CONCEPT FOR SUBGRADE MODELS

4.3.1 Background and Overview

There are numerous subgrade models that have been developed over the course of approximately 200 years. At first glance, they are seemingly based on multiple physical visualizations (some of which yield the same mathematical outcome although they are visually very different) and concepts that have no single, common element. In some instances, different names have been used for the same subgrade model. In other instances, different names and notation have been used for the same parameter in a subgrade model. All of this collectively results in a complex and confusing picture to potential users of subgrade models.

A central element of this monograph in general, and this chapter in particular, is to illustrate that all subgrade models are, in reality, approximations of the same theoretical problem, that of a homogeneous, isotropic, linear-elastic continuum with an applied normal stress on its surface. The basic elements of this unifying concept have been known for over 50 years (Rhines 1965, Kerr and Rhines 1967) although for any number of reasons this knowledge has never been adequately transmitted to and disseminated among foundation engineers or even academicians involved in foundation engineering research.

As it turns out, one key reason for the perception that subgrade models as a group have no common element is that there are two distinctly different ways in which to physically visualize this unifying elastic continuum and thus develop the subgrade models that approximate its stress-displacement behavior:

- by applying arbitrary simplifications to the partial differential equations governing the behavior of the original elastic continuum to produce what are hereinafter referred to as *simplified-continuum* subgrade models and

- by physically replacing the original elastic continuum with some arbitrary assemblage of 'mechanical' elements such as independent axial springs, deformed tensioned membranes, shear-only beams/plates, and flexure-only beams/plates that produce what are hereinafter referred to as *mechanical* subgrade models[28].

A logical question at this point is why do mechanical subgrade models even exist no less are used as opposed to the former simplified-continuum models. Stated another way, why develop and pursue a contrived, abstract visualization using mechanical elements of a physical object (an elastic continuum in this case) whose actual, inherent visualization is obvious and relatively straightforward to work with directly.

The answer is that the latter mechanical visualization actually occurred over a century <u>before</u> the former simplified-continuum concept was conceived of and published. Thus, an abstract physical system using mechanical elements was postulated, developed mathematically, and widely used in both practice and research before anyone realized that it represented an actual physical system that could be approximated directly.

There have also been significant shortcomings with regard to technology transfer to foundation engineers who actually use subgrade models. Published work related to the simplified-continuum concept for subgrade models has been a fraction of that devoted to the traditional mechanical subgrade models. Furthermore, published work related to simplified-continuum subgrade models has generally been limited to publication venues directed at applied mechanicians and mechanical engineers, not civil engineers. Thus, despite their existence for over 50 years, subgrade models based on the simplified-continuum concept have not appeared in basic geotechnical engineering textbooks that by their nature are the fundamental link of one generation of foundation engineers to the next.

In any event, this chapter will not only show how simplified-continuum and mechanical subgrade models are solutions for the same basic problem but will combine and synthesize them into a single, hybrid concept for subgrade models that combines the best elements of each into a new, practical, practice-oriented approach to subgrade modeling using commercially available structural analysis software.

As a final comment, for the sake of completeness it is noted that Kerr and Rhines (1967) make brief mention of a third conceptual approach for developing subgrade models. This is a mathematical approach involving integral equations for stress-displacement (similar to what are used for the BEM such as discussed by Fraser and Wardle (1976)) and

---

[28] The SSI application of a horizontal foundation element supported on the simplest mechanical subgrade model consisting of a system of independent axial springs is often referred to in the literature as the *Beam on Elastic Foundation* (BOEF) problem. Furthermore, an analysis performed using this model is often referred to in older literature (e.g. DeSimone and Gould 1972) in particular as an 'elastic' analysis. The origin of the BOEF term is unclear to the writer but references to this spring model as an 'elastic foundation' appear in published technical papers in the applied mechanics literature as early as the 1930s (e.g. Biot 1937). Furthermore, Biot implies that this terminology had already been in use for some time by then as the spring foundation was viewed as an approximation of a 'true' elastic-continuum subgrade. What is clear to the writer, however, is that the complete BOEF term was made forever famous by virtue of its use as the title of Hetényi's seminal scholarly monograph on the subject of subgrade modeling and models (Hetényi 1946). In any event, it is essential to note that the use of the term 'elastic' in this context is coincidental and unfortunate for the confusion it creates as it has absolutely nothing to do with the theory of elasticity; a true elastic continuum; or the fact that the theory of elasticity turns out to be the unifying concept for all subgrade models regardless of their developmental approach, i.e. as a simplified elastic continuum or assemblage of mechanical elements. Because of the longstanding use of the BOEF term, there is realistically no chance that this confusing, incorrect use of the term 'elastic' will ever change in the engineering lexicon. However, as a small contribution toward deprecating this confusing terminology, the term BOEF will not be used further in this monograph.

then retaining only what Kerr and Rhines call "low-order" terms of those equations. The several references dealing with this third conceptual approach that are cited in Kerr and Rhines (1967) are primarily in the French and German languages. The writer has not investigated subgrade models using this third approach and they are not addressed further in this monograph.

4.3.2 Early Research

Interest in developing solutions for SSI problems involving a true linear-elastic continuum as a subgrade model date back to at least the early decades of the 20th century in the English-language, applied-mechanics literature. For example, Biot (1937) considered the problem of an infinitely long elastic beam subjected to a point load and supported on an isotropic, homogeneous, elastic half space. Bosson (1939, a more-accessible reference is Hetényi (1946) who reproduced Bosson's work) considered the plane-stress problem of an elastic beam with a rectangular cross-section supported on an elastic continuum of finite width (the same as that of the beam) and finite thickness. Several decades later, Vesic and Saxena (1970) considered the problem of an elastic plate subjected to a point load (their interest was primarily PCC-pavement related) that was resting on an elastic continuum.

While these and other publications illustrated that rigorous solutions for SSI applications involving an elastic continuum were achievable, no outcomes of broad utility for developing subgrade models explicitly based on an elastic continuum are obvious from this body of work. This is because all of the cited references involved the inherent existence of some type of foundation element (beam or plate) supported on the elastic continuum. The presence and explicit inclusion of a foundation element obscures the stress-displacement relationship of the continuum itself because the theoretical results are inherently dependent on and involve the flexural stiffness of the foundation element. Indeed, the purpose of each of these studies was not directed toward subgrade modeling and models in general but to providing what was assumed to be a theoretically rigorous benchmark solution against which results from a comparable problem of a beam or plate supported on an inherently simpler, approximate (but easier to use in practice) subgrade model could be compared. Keep in mind that the vast majority of this cited research was conducted in the era prior to the availability of digital computers and numerical solutions using same, so rigorous, closed-form solutions involving the theory of elasticity were the 'gold standard' for use as a comparative benchmark for evaluating other, simpler subgrade models.

4.3.3 Rhines' Solution

In the writer's opinion, credit for the real research breakthrough for illustrating that a theory-of-elasticity solution is the unifying basis for all subgrade models regardless of how they are conceived and developed belongs to Dr. Warren J. Rhines and, indirectly, his mentor and doctoral adviser at the original New York University engineering school, the late Prof. Arnold D. Kerr. They collectively made what the writer feels are seminal and crucial contributions to the knowledge base for subgrade models that are largely unknown to most civil engineers as their published works were primarily in applied-mechanics venues or in the form of research reports submitted to funding agencies. The numerous contributions of Kerr and Rhines to the overall subject of subgrade models will become clear later in this chapter but at this point the discussion is limited to one element of Rhines' doctoral dissertation (Rhines 1965) that was also summarized in Kerr and Rhines (1967).

Rhines solved a problem in linear elasticity that is broadly similar to Bosson's plane-stress problem but without the foundation element on the surface of the elastic continuum. Furthermore, Rhines assumed plane-strain conditions so that surface settlements, w, were only functions of one horizontal direction, x. Specifically, Rhines considered a vertical stress, p, applied directly to an elastic continuum of finite thickness, H, that was underlain by a rigid base.

Rhines expressed the solution to this problem as the equivalence of two infinite series involving total derivatives in x of the applied stress, p, in one series and the surface settlement, w, in the other series. The same solution was shown in several different formats in Rhines (1965) and Kerr and Rhines (1967) depending on how the authors chose to move the constant coefficients from one series to the other in order to illustrate different aspects of the solution.

The overall form of Rhines' solution of greatest relevance to this monograph is (note that total derivatives are shown in the usual abbreviated format as superscripts):

$$p^{II} - C_{p1}p^{IV} + C_{p2}p^{VI} - \cdots = C_{w1}w^{II} - C_{w2}w^{IV} + C_{w3}w^{VI} - \cdots \tag{4.1}$$

where C_{pi} and C_{wi} are constants that involve combinations of only the thickness of the elastic continuum, H, and the continuum elastic parameters E and ν.

The real conceptual power of Equation 4.1 is that Rhines (1965) and Kerr and Rhines (1967) showed that all subgrade models developed using either the simplified-continuum or mechanical approaches noted previously and that are discussed in detail later in this chapter have the exact same form as Equation 4.1 after integrating both sides of Equation 4.1 twice and evaluating the integration constants. The only difference in the outcomes is that for mechanical subgrade models the equation coefficients C_{pi} and C_{wi} are functions of the parameters that define the behavior of the mechanical elements, e.g. spring stiffness, k, whereas for simplified-continuum subgrade models the equation coefficients involve H, E, and ν.

In the writer's opinion, the work and observations of A. D. Kerr and his protege Rhines is nothing short of extraordinary in its implications for subgrade models as it provides the theoretical basis and desired unifying key to relating all subgrade models, whether derived using the simplified-continuum approach or mechanical approach, to what is arguably the simplest of all 'complete' soil models, that of a linear-elastic continuum. This is because any solution to an infinite series that retains only a finite number of terms of that series (not necessarily the same number of terms on each side of the equation) is always an approximate solution to the exact solution.

Furthermore, because the simplified-continuum and mechanical approaches to developing subgrade models produce equations of the same basic form and differing only in how the equation constants C_{pi} and C_{wi} are defined, this provides the underlying groundwork for developing what the writer terms *hybrid* subgrade models wherein unique relationships between the parameters defining mechanical elements in a subgrade model and elastic parameters of an actual subgrade can easily be developed. The practical consequence of this is that hybrid subgrade models can be readily implemented into commercially available structural analysis software (by virtue of their mechanical elements) yet the constant coefficients associated with those mechanical elements can be rigorously evaluated using fundamental subgrade properties determined by project-specific site characterization by virtue of the relationships involving elastic parameters. This entire process is illustrated for several case histories in Chapters 5 and 6.

In summary, the unifying concepts for all subgrade models that are drawn from the Rhines problem in the theory of elasticity are that:

- a linear-elastic continuum can be viewed as the simplest of all 'complete' soil models;

- one way of expressing the solution to a basic elastic-continuum problem of an applied normal stress, p, on the surface of the continuum and the settlement, w, of the continuum surface is as an equality of two infinite series involving only derivatives of these two variables and the geometry and conventional material parameters of the elastic continuum;

- any solution to this elastic-continuum problem that contains only a finite number of terms in the two infinite series can be viewed as an approximate solution to the problem, with the overall approximation increasing in accuracy with an increasing number of retained terms;

- every subgrade model identified to date and regardless of how it was arrived at can be expressed as a finite number of terms involving derivatives of p and w; and

- the conclusion that can be drawn is that every subgrade model identified to date can thus be viewed as an approximation to the same theory-of-elasticity problem.

4.4 ORGANIZATION

4.4.1 General Overview

The remainder of this chapter presents the details to support the unifying hypothesis that was outlined in the preceding section. The only question is how to best organize the presentation of what amounts to approximately 200 years of published work.

The writer's earlier published works that included a detailed history and discussion of the evolution of subgrade models (Horvath 1979, 1988d, 1989c, 2002) were organized in a logical order of temporal progression beginning with the genesis work attributed by early researchers to Fuss and Winkler in the 19th century. However, as noted in Chapter 1, this monograph embraces a different organizational approach in an effort to better harmonize and ultimately synthesize the disparate approaches to subgrade modeling that have occurred over the last 200 years. This is done primarily with the intention of trying to achieve greater overall unity and clarity in the presentation.

Nevertheless, for reasons that will become clear it is still desirable to begin the discussion with the Fuss-Winkler Hypothesis. However, the initial discussion is limited to the original abstract definition of this hypothesis which, strictly speaking, is an outlier that is neither a simplified-continuum nor mechanical type of subgrade model as no physical model was apparently postulated by either Fuss or Winkler to complement and illustrate their mathematical abstraction.

The simplified-continuum category of subgrade models is then discussed as these models more clearly and fundamentally illustrate the precept put forth earlier in this chapter that all subgrade models are approximations of the same, basic theory-of-elasticity problem. This is followed by a discussion of mechanical subgrade models, including the long-standing, ubiquitous visualization of the Fuss-Winkler Hypothesis using mechanical elements (which is why it is both pragmatic and convenient to categorize the otherwise unique Fuss-Winkler Hypothesis as a mechanical subgrade model as the writer has consistently done in earlier published works).

This presentation of the two fundamental types of subgrade models is followed by a discussion of the newly defined category of hybrid subgrade models that combine the best elements of each of the two traditional categories of subgrade models. As will be seen, hybrid subgrade models represent the current state of art for subgrade modeling and examples of the use of hybrid subgrade models in practice are the focus of Chapters 5 and 6.

4.4.2 Additional Comments

Before beginning the presentation of the remainder of this chapter, two additional comments are in order. First, the discussion of simplified-continuum, mechanical, and hybrid subgrade models in the sections immediately following presumes their use to represent the response of only the subgrade component of an SSI application involving some foundation element in contact with a subgrade. This is not a trivial statement as recent decades have seen a modest body of published work in which a single subgrade model was used to represent a composite subgrade response, typically a soil subgrade reinforced with a horizontal layer or layers of a traditional, planar geosynthetic such as a geogrid, geomembrane, or geotextile. In such cases, the subgrade model reflects the behavior of not only the subgrade soil but the tensioned geosynthetic as well. Such composite applications of subgrade models are discussed in a separate section at the end of this chapter as the subgrade-model parameters have very different interpretations compared to traditional subgrade-only uses of subgrade models.

The second comment involves an analytical methodology that the writer calls the *Conventional Method of Static Equilibrium* (CMSE). This methodology has long been used for SSI applications, especially mat foundations and related shallow foundations such as combined footings and especially in the pre-computer era when analytical alternatives were far more limited than they are at present. Although the CMSE is frequently included in a discussion of subgrade models (including by the writer in earlier publications) because it provides an estimate of the subgrade reaction, p, it is not considered a subgrade model in this monograph. This is because the CMSE can only be used to calculate bending moments in a foundation element and in the process completely ignores the effect of relative foundation-subgrade flexibility (the essence of SSI and subgrade models) on those moments. However, because the CMSE has historically played such a significant, outsize role in SSI applications as a de facto subgrade model, especially with mat foundations, it does warrant some discussion in this monograph. That discussion is presented in Appendix B.

4.5 THE FUSS-WINKLER HYPOTHESIS

The Fuss-Winkler Hypothesis is the cornerstone of subgrade modeling and models in SSI analyses, if for no other reasons than by virtue of its age (currently of the order of 200 years) that precedes all other known attempts to develop subgrade models as well as its remarkable endurance and persistence in both practice and research despite its well-known analytical shortcomings. However, there are a number of misunderstandings about the Fuss-Winkler Hypothesis that persist in the state of knowledge to the present and require clarification.

To begin with, using the geotechnical sign convention as defined in this monograph and illustrated in Figure 2.1, the Fuss-Winkler Hypothesis is stated as:

$$p(x,y) = k_{FW}(x,y) \cdot w(x,y) \tag{4.2}$$

where $k_{FW}(x,y)$...abbreviated as k_{FW} hereinafter for simplicity...is the *Fuss-Winkler coefficient of subgrade reaction.*

To being with, Equation 4.2 is not simply a rearranged version of Equation 3.1 that defined the generic coefficient of subgrade reaction, k. There is a very significant difference between Equations 3.1 and 4.2 that is not apparent from simply the arithmetic expressions. Specifically, the former equation (3.1) defines k as an arbitrary[29] parameter that is the calculated outcome of two calculated outcomes from an ideal SSI analysis as shown in Figure 3.2. On the other hand, the latter equation (4.2) uses k_{FW} as a necessary input variable for defining subgrade reaction, p, for use with several of the 'non-ideal' analytical methodologies defined in Chapter 3 that are pragmatic alternatives to the ideal solution alternative and require the subgrade reaction to be input and not calculated as in the ideal alternative. Note that this includes the structural alternative that was identified as the alternative that is currently considered the optimal one to use in most SSI applications, especially mat foundations. Thus, the Fuss-Winkler Hypothesis reflects the tectonic paradigm shift with regard to the nature of the coefficient of subgrade reaction, k, from incidental calculated outcome to necessary input parameter that was discussed at length in Chapter 3.

There are two fundamental issues to note at the outset with respect to the Fuss-Winkler Hypothesis:

- There is an inherent error that simply can never be corrected. Specifically, the Fuss-Winkler Hypothesis does not replicate the settlement behavior of actual foundations as shown idealistically in Figure 3.4 to the extent that there will be settlement outside of a loaded area, i.e. where $p = 0$.

- All subgrade behavior is folded into a single parameter, k_{FW}. This turns out to be the single biggest shortcoming of the Fuss-Winkler Hypothesis. As will be seen later in this chapter, at least two independent parameters are required in order for a subgrade model to inherently be able to replicate the most basic elements of actual subgrade behavior. It is possible in principle to overcome this requirement, at least to a significant extent, when using only one parameter as with the Fuss-Winkler Hypothesis. However, to do so places a unique burden on that sole parameter (k_{FW}) and, most importantly, on the foundation engineer. Specifically, it requires the correct answer to be known beforehand so that the correct values of k_{FW} can be selected in order to produce the correct answer. While this statement sounds facetious and impossible to satisfy, it is nonetheless a statement of fact. Rather remarkably, it will be seen that in at least one very narrow, well-defined SSI application (laterally loaded deep foundations) it is actually possible to do this routinely.

There are several other significant aspects of the Fuss-Winkler Hypothesis that require mention and discussion in order to set the record straight and hopefully eliminate some of the aforementioned misunderstandings that continue to permeate the state of knowledge to the present:

- The Fuss-Winkler Hypothesis is an abstract concept and a mathematical statement of assumption. No physical model of what k_{FW} might represent was expressed or implied.

---

[29] In this context, 'arbitrary' means that this variable is not necessary to interpret the outcomes of an ideal analysis as both the calculated subgrade reaction, p, and calculated settlement, w, are unambiguous in and of themselves and thus do not require the definition of k to facilitate their interpretation. Rather, k is just an arbitrary parameter that is useful for defining how p and w are related at a given point.

Common physical visualizations involving a layer of independent axial springs and body of liquid that are discussed later in this chapter evolved sometime later and are not necessary for its use.

- There was no a priori assumption or limitation imposed on the relative spatial variation of k_{FW}. Specifically, there was no requirement for it to be uniform in magnitude in any given problem. In fact, as can be seen with the idealized limiting cases shown in Figure 3.4 it will never be uniform in magnitude. Rather, simplistic assumptions concerning spatial variation that were made subsequently by various researchers were dictated by the pre-computer era when closed-form solutions in the form of algebraic expressions were the most theoretical outcome that could be achieved. In order to achieve such solutions, relatively simple variations in k_{FW} had to be assumed as is well-illustrated in what is arguably the most complete English-language treatment of the Fuss-Winkler Hypothesis, Hetényi's seminal 1946 scholarly monograph. While Hetényi developed most solutions assuming k_{FW} = constant, he also included a solution where k_{FW} varied in linearly in one dimension.

- Use of the Fuss-Winkler Hypothesis is not limited to geotechnical/SSI applications. In fact, Hetényi (1946) noted on the very first page of Chapter 1 of his monograph that the hypothesis is actually a much more accurate analytical model for several purely structural applications. Note that Hetényi was not a geotechnical specialist per se but a specialist in the broader field of applied/engineering mechanics that historically has applications to not only civil engineering but mechanical engineering as well.

- The limitations and approximations of the Fuss-Winkler Hypothesis as a model for soil subgrades were acknowledged by Hetényi (1946). However, given the very limited alternatives (primarily closed-form solutions for relatively simple problems involving a linear-elastic continuum which Hetényi also discussed in some detail) in that pre-computer era, the Fuss-Winkler Hypothesis was viewed as a reasonable, pragmatic compromise for use.

However, one of the most intriguing and relevant issues to note about the Fuss-Winkler Hypothesis is found in published work by Dr. Karl von Terzaghi (who went by the surname of simply Terzaghi in later years) whose larger-than-life stature in the world of all things geotechnical persists (although diminishing slowly as time goes on) to the present, especially in textbooks and design handbooks.

Terzaghi authored what appears to be the first detailed English-language publication that focused on actual geotechnical/foundation engineering use of the Fuss-Winkler Hypothesis in general and ways to evaluate k_{FW} in actual project applications in particular (Terzaghi 1955)[30]. Although elements of his 1955 paper (primarily the formulas and tables suggesting how to evaluate k_{FW}) can still be found in many current publications, significant textual elements of his 1955 paper seem to have been forgotten over time. As a result, Terzaghi's recommendations concerning the evaluation of k_{FW} tend to be used out of context.

---

[30] It is of interest and relevance to note that this was not Terzaghi's first English-language published work to mention the subject of subgrade models. He addressed the topic to a limited extent using a totally different, more-theoretical approach in his first English-language textbook (Terzaghi 1943). Curiously, he did not mention his earlier work at all in his 1955 paper. Furthermore, subsequent publications by others that discuss subgrade modeling and models generally only mention Terzaghi's 1955 treatment of the subject, with no mention of his 1943 thoughts on the matter.

56

Specifically and most importantly, Terzaghi did not view the Fuss-Winkler Hypothesis as a subgrade model in the context of this monograph, i.e. as an analytical tool to use for accurately forecasting both bending moments and displacements (settlements in most cases) in the foundation element interacting with the ground. This is rather clear in this quote taken from Terzaghi (1955):

> "...the theories of subgrade reaction should not be used for the purpose of estimating settlement or displacements."

Rather, Terzaghi saw the Fuss-Winkler Hypothesis solely as a useful <u>structural engineering</u> tool...and a rather limited one at that...to assist in forecasting values of bending moments that were more realistic than those from using the CMSE that was noted earlier in this chapter and is discussed in detail in Appendix B. Terzaghi noted that structural elements in SSI applications generally exhibit some flexibility. As is well known in structural mechanics, all other things being equal, bending moments tend to decrease as structural flexibility increases (and its reciprocal, rigidity, decreases). With smaller bending moments, less substantial and less costly structural members are required. Of course, the 'price' that is paid for the reduced bending moments and smaller structural members is increased displacement and deformation of the structural element or system. This is the challenge faced by structural engineers in many applications, striking a cost-effective balance between structural flexibility/rigidity and displacement/deformation.

In any event, Terzaghi viewed the bending moments in, say, a mat foundation as having two components or contributions:

- *local* moments due to the relatively localized effect of a concentrated load from, say, a building column (this relates to the parameters radius of stiffness, r_0, and range of influence, R, that were discussed in Chapter 3) and

- *global* moments due to the *sagging* a.k.a. *dishing* mode (pattern) of overall settlement of a mat due to the cumulative, overlapping vertical-stress effect on the underlying subgrade of all applied loads.

Note that in this vision of bending-moment development, only the relatively shallow subgrade immediately below the foundation element (mat in this case) influences the local moments whereas it is the deeper portions of the subgrade that contribute to the global moments. In any event, Terzaghi's recommendations for evaluating k_{FW} were developed and framed with the narrow-intended use for <u>calculating local moments only</u>.

Setting aside for a minute Terzaghi's admonition not to use the Fuss-Winkler Hypothesis as a settlement-forecasting tool, experience in practice has long been that calculations made using Terzaghi's 1955 recommendations for k_{FW} typically produce very small values of settlement (fraction of one inch or a few millimetres) that are generally an order of magnitude smaller than the actual settlements. This fact was noted in Chapter 3 in the discussion of the traditional solution alternative that historically uses the Fuss-Winkler Hypothesis as the subgrade model for analysis of the mat component. This is also the reason why it was noted in Figure 3.3 for this analytical alternative that the values of subgrade reaction assumed for bending-moment analysis of the mat component are generally different from the subgrade reaction used in the settlement analysis of the subgrade component. In order to produce realistic settlements for the geotechnical component of the overall problem, design professionals typically assumed a subgrade reaction $p'(x,y)$ that is very different from the subgrade reaction $p(x,z)$ that is used for the mat component. Consistent with Terzaghi's

1955 recommendations, there is no unity or compatibility between subgrade reactions and settlements of these two analytical components.

One additional observation made in Terzaghi (1955) is relevant to note and that is that calculated values of bending moments are relatively insensitive to the value of k_{FW} that is used. This is simply a happenstance of structural mechanics. On the other hand, calculated values of displacement are much more sensitive to the value of k_{FW} as can easily be seen in the one-to-one relationship between k_{FW} and displacement implied by Equation 4.2.

4.6 SUBGRADE MODELS BASED ON THE THEORY OF LINEAR ELASTICITY

4.6.1 Reissner's Simplified-Continuum Concept and Solution

The late Prof. Max(well) Erich 'Eric' Reissner made what the writer considers to be the defining contribution to date to the simplified-continuum approach to developing subgrade models. Reissner considered the problem of a linear-elastic layer of infinite horizontal extent but finite thickness, H, and underlain by a rigid base. The elastic parameters of the layer, E and G, were assumed to be depth-wise constant. A normal stress of magnitude $p(x,y)$ was assumed to be applied to the layer surface and cause surface settlements $w(x,y)$. For simplicity, these parameters are denoted as p and w in the following presentation.

It is significant to note in view of future use of Reissner's work in this monograph that the finite thickness of the elastic layer and its underlying rigid base can be virtual and not explicit in real-world applications. From an analytical perspective, the depth, H, to the rigid base need only be a depth beyond which contribution to problem settlements can be reasonably taken to be zero.

The starting point for developing Reissner's concept for subgrade models is recognizing that the behavior of an elastic continuum is governed by three sets of partial-differential equations:

- constitutive (stress-strain),

- strain-displacement, and

- equilibrium.

A complete, rigorous solution requires consideration and satisfaction of all of these equations plus boundary conditions. Reissner suggested that by making judicious assumptions concerning various stress, strain, and displacement parameters (i.e. that they be assumed to be equal to zero or are otherwise defined beforehand) the overall solution process could be greatly simplified, albeit yielding a solution that is only approximate.

The specific assumptions made in Reissner's original paper on the subject (Reissner 1958) were that the horizontal normal and shear stresses within the elastic layer were assumed to be everywhere zero. Furthermore, the horizontal normal displacements were assumed to be zero at the rigid base that defined the bottom of the layer. The horizontal normal displacements at the surface of the layer had to be consistent with an assumed boundary condition that is discussed further below. Reissner also neglected body forces (overburden stresses in a geotechnical context) and compatibility between stresses and displacements.

The writer (Horvath 1979, 1983a with important corrections in 1984b) rederived Reissner's original problem using the geotechnical sign convention shown in Figure 2.1.

Reissner used an atypical coordinate system that had its origin at the bottom of the elastic layer and was not conducive for use in foundation applications. The resulting partial-differential equation relating the applied stress, p, and surface settlement, w, for the limiting case of zero horizontal displacements on the surface of the elastic layer is:

$$p - \left(\frac{GH^2}{12E}\right)\nabla^2 p = \left(\frac{E}{H}\right)w - \left(\frac{GH}{3}\right)\nabla^2 w \,. \tag{4.3}$$

For the sake of completeness, it is noted that the writer also solved what will hereinafter be referred to as the *Reissner Simplified Continuum* (RSC) subgrade model for the cases of E and G increasing linearly with depth[31] and with the square root of depth, in both cases starting from an arbitrary non-negative value at the surface. The solutions, which were presented only in Horvath (1979), have the same basic form as Equation 4.3 but with constant coefficients that are substantially more complex algebraically. Given the way that subgrade models have evolved into the hybrid formulation discussed later in this chapter and used as illustrated in Chapter 6, these two additional solutions never found any practical use.

Some additional observations concerning the RSC are in order, also for the sake of completeness. To begin with, Reissner allowed for a time-dependent (viscoelastic) response of the elastic layer. This results in the entire right-hand side of Equation 4.3 being multiplied by the term:

$$\left(1 + \lambda \frac{\partial}{\partial t}\right) \tag{4.4}$$

where λ is an arbitrary constant. Reissner did not explore any solutions or examples or identify any practical applications/consequences of time-dependent behavior in either his original (Reissner 1958) or subsequent (Reissner 1967) papers concerning the RSC. However, others (the aforementioned A. D. Kerr in particular) explored viscoelastic behavior of mechanical subgrade models in some detail as discussed later in this chapter.

Reissner also considered the presence of an elastic plate with an applied vertical stress $q(x,y)$ that was supported on the surface of the elastic layer in both his 1958 and 1967 papers. In this case, p becomes the subgrade reaction beneath the plate and the problem is analogous to what was illustrated in Figure 2.4 for a beam.

Assuming thin-plate behavior as defined in Chapter 2, the governing partial-differential equation for an elastic plate of flexural stiffness D with a generic subgrade reaction defined by p is:

$$\nabla^2(D\nabla^2 w) + p = q \tag{4.5}$$

that simplifies to:

$$D\nabla^4 w + p = q \tag{4.6}$$

if D is constant. Either Equation 4.5 or 4.6 can be rearranged in terms of p and then integrated into Equation 4.3 to create a single sixth-order partial-differential equation that defines the coupled behavior of the plate-RSC subgrade system.

---

[31] This is often referred to in the literature as a *Gibson soil of the first kind* (the second kind being Gibson's rarely noted hyperbolic variation with depth) or, more commonly, simply a *Gibson soil*.

The presence of an elastic plate in Reissner's problem allows for broadening the boundary-condition assumption on horizontal displacements at the surface of the elastic layer. One limiting case is that of zero displacement and this can be viewed as the perfectly smooth limiting case with a plate. The other limiting case then becomes the perfectly rough case (Pister and Williams 1960) which is the same as saying that the horizontal displacement of the surface of the elastic layer must match the horizontal displacement of the bottom of the plate in its deformed position.

The writer (Horvath 1979, with the final result only in Horvath and Colasanti 2011a) derived the governing equation of an RSC with this perfectly rough boundary condition which is as follows:

$$p - \left(\frac{GH^2}{12E}\right)\nabla^2 p = \left(\frac{E}{H}\right)w - \left[\left(\frac{GH}{2}\right)\left(\frac{2}{3} - \frac{t}{2H}\right)\right]\nabla^2 w \,. \tag{4.7}$$

In any practical application, the thickness of the plate, t, relative to $2H$ will be very small so the difference relative to the perfectly smooth case (Equation 4.3) is also very small.

Of much greater importance and necessity in any practical use of the RSC as a subgrade model is the issue of subgrade boundary conditions at the edge of the stress, p, that is applied to the subgrade which is the same as the edge of the plate if one is present. This is because when $p = 0$ the same is not true of w as can be seen by inspection of Equations 4.3 and 4.7. Reissner touched on the subject of subgrade boundary conditions briefly in his 1958 paper but discussed it in greater detail in his 1967 paper as the latter was essentially a rebuttal to two specific publications by the aforementioned W. J. Rhines concerning Reissner's work (Rhines 1966, 1967).

The subject of subgrade boundary conditions requires detailed discussion as it also comes up with almost all other subgrade models based not only on simplified elastic continuums but mechanical elements as well. Consequently, a general discussion of this topic is presented later in this chapter.

4.6.2 Extension of Reissner's Simplified-Continuum Concept

Reissner (1958) made an indirect suggestion that the concept of simplifying the aforementioned suite of partial-differential equations that govern the behavior of an elastic continuum in order to create a subgrade model could be extended beyond the specific assumptions he made. In particular, he noted that the form of the Fuss-Winkler Hypothesis could easily be obtained by setting G in Equation 4.3 equal to zero (equivalent to an elastic layer with no shear stiffness in any direction). Subsequently, Rhines (1965) and Kerr and Rhines (1967) made much more formal and explicit suggestions in this regard.

As part of doctoral research in the 1970s, the writer pursued the formal development of subgrade models using Reissner's basic concepts for a simplified elastic continuum and produced two models simpler than the RSC as a result. The first model added the assumption that horizontal displacements within the elastic layer were everywhere zero to the previous assumptions made by Reissner as had been suggested by Rhines (1965) and Kerr and Rhines (1967). As with the RSC, the writer developed solutions assuming that the elastic parameters of the layer were uniform with depth as well as increased linearly and with the square root of depth, beginning from an arbitrary non-negative value at the surface in both cases. The detailed results were presented in Horvath (1979) but only the final governing equation for the basic case of depth-wise constant elastic parameter is presented here (it was also published in Horvath 1988d, 1989c, 2002):

$$p = \left(\frac{E}{H}\right)w - \left(\frac{GH}{2}\right)\nabla^2 w \ . \qquad \textbf{(4.8)}$$

For reasons that will become clear later in this chapter, the writer named this the *Pasternak-Type Simplified Continuum* (PTSC). Note that subgrade boundary conditions at the edge of the loaded area need to be addressed with the PTSC as they did with the RSC.

The second simplified-continuum subgrade model developed by the writer formalized what Reissner had suggested in 1958. By adding the assumptions that the vertical shear stresses were zero to all prior assumptions, the following governing equation was obtained for the basic case of the elastic layer parameters being depth-wise constant:

$$p = \left(\frac{E}{H}\right)w \ . \qquad \textbf{(4.9)}$$

For obvious reasons (compare Equation 4.9 to Equation 4.2), the writer named this the *Winkler-Type Simplified Continuum* (WTSC) although in keeping with the conventions adopted in this monograph it is herein and hereinafter renamed the *Fuss/Winkler-Type Simplified Continuum* (F/WTSC).

Detailed derivation of the F/WTSC for three different assumptions concerning the depth-wise variation of the elastic-layer parameters was presented in Horvath (1979, 1983b with important corrections in 1984c). Just the final result for depth-wise constant values of the elastic parameters as shown in Equation 4.9 was also presented in Horvath (1988d, 1989c, 2002).

As an aside, note that there is no subgrade boundary-condition issue at the edge of the loaded area with the F/WTSC as there is with both the RSC and PTSC. In Equation 4.9, it is clear that when $p = 0$ that $w = 0$ also.

Table 4.1 contains a summary of the assumptions made in deriving the governing equations for the RSC, PTSC, and F/WTSC. This is to facilitate comparison between and among the different subgrade models derived using Reissner's concept for simplified continuums. In addition and as noted previously, all simplified-continuum subgrade models developed using Reissner's conceptual approach neglect body forces and compatibility between stresses and displacements within the elastic layer.

Table 4.1. Assumptions Made for Reissner-Type Simplified-Continuum Subgrade Models.

Parameters		Subgrade Model		
		RSC	PTSC	F/WTSC
normal stresses	σ_z	variable	variable	variable
	σ_x, σ_y	0	0	0
shear stresses	τ_{xz}, τ_{yz}	variable	variable	0
	τ_{xy}	0	0	0
displacements	w	variable	variable	variable
	u, v	(see note)	0	0

Note: Value at top (surface) of elastic layer depends on perfectly smooth vs. perfectly rough boundary condition chosen, otherwise variable within elastic layer and zero at bottom of elastic layer.

Of greatest interest in this tabulation is that the most significant theoretical shortcoming of the F/WTSC and, by extension, the Fuss-Winkler Hypothesis itself, is that vertical shear stresses within the elastic layer (= subgrade) are neglected. Shear resistance of soil in the direction of load application is a fundamental behavioral mechanism for load resistance. Furthermore, shear resistance is the only fundamental strength mechanism for soil so is a key load-transfer mechanism within the subgrade in SSI applications. Thus, to neglect shear resistance completely as the Fuss-Winkler Hypothesis implies and all subgrade models such as the F/WTSC that are related to it do results in a significant behavioral shortfall.

Of course, this shortcoming of the Fuss-Winkler Hypothesis has been known for the better part of 100 years so is nothing new. The simplified-continuum concept for subgrade models as summarized in Table 4.1 simply formalizes that knowledge within the well-known theoretical framework of a linear-elastic continuum. This table also indicates that by increasing the theoretical rigor of the F/WTSC subgrade model by even one mathematical level (i.e. to the PTSC) results in significant improvement as vertical shear stresses are now explicitly included in the subgrade model.

As a final comment, the discussion up to this point has focused on starting with the RSC and making further simplifying assumptions to develop even simpler subgrade models. It is clear from Table 4.1 that one could go in the opposite direction, i.e. starting with the RSC and developing more-exact subgrade models by either relaxing one or more of the assumptions made concerning the horizontal normal and/or shear stresses or by considering stress-displacement compatibility or both. Some years ago, the writer began preliminary research along these lines but did not progress to any usable conclusion. Consequently, this research path remains to be pursued and developed.

4.6.3 The Simplified-Continuum Concept and Solutions of Vlasov and Leont'ev

Reissner was not alone in his approach to developing subgrade models by explicitly simplifying the equations that define an elastic continuum. The writer (Horvath 1979, 1988d with a summary only in 1989c) is aware of at least one more conceptual approach along these lines that was undertaken by Vlasov and Leont'ev (1960)[32]. They achieved a solution by using a combination of variational calculus (an approach unique to them) together with explicit assumptions concerning certain elastic parameters being equal to zero (as did Reissner).

The writer has identified two versions of Vlasov and Leont'ev's work. They are referred to as the "one-layer model" and "two-layer model" in prior published works by the writer that include tabular summaries of subgrade models (Horvath 2002, Horvath and Colasanti 2011a). Those labels are retained in this monograph.

The overall most significant and distinctive element of the Vlasov-Leont'ev analytical methodology in both cases is that they assumed and defined the depth-wise variation of the settlement of the elastic layer, w, as part of the problem input as opposed to allowing w to be an outcome of the solution process as did Reissner in the RSC (and the writer in the PTSC and F/WTSC extensions of Reissner's methodology summarized in the preceding section). Specifically, w is defined using a function Vlasov and Leont'ev called $\phi(z)$ that includes a dimensionless constant they called γ. Note that these two parameters bear no relation to the usual geotechnical definitions of ϕ (Mohr-Coulomb strength parameter) and γ (soil unit weight).

---

[32] In some publications, including some earlier published works by the writer, the date of this reference is given as 1966. Note that the original publication of this document in the Russian language appeared in 1960 while the publication of an English-language translation did not appear until 1966. Hence, the dual dates. In addition, in most English-language publications by others, Leont'ev is spelled Leontiev. However, the former spelling is used in this monograph.

Although the concept used by Vlasov and Leont'ev was expanded on by other applied-mechanics researchers (e.g. Kamesovara Rao et al. 1971) and the subgrade models resulting from Vlasov and Leont'ev's work received some modest attention in geotechnical publications several decades ago (e.g. Harr et al. 1969, Scott 1981), the writer does not consider the Vlasov-Leont'ev models to be anything other than theoretical curiosities. This is because their one- and two-layer model outcomes are no more accurate than the PTSC and RSC, respectively, in terms of the overall form of their governing equations plus require the a priori assumptions involving the aforementioned parameters ϕ and γ. Thus, the work of Vlasov and Leont'ev is mentioned here solely for the sake of completeness and to illustrate that Reissner was not alone in pursuing subgrade models by simplifying linear-elastic continuums.

As a final comment, note that the issue of boundary conditions beyond the limits of the loaded area arises with the Vlasov-Leont'ev models as it does with both the RSC and PTSC.

4.6.4 Direct Application of Rhines' Solution

Both Reissner and Vlasov-Leont'ev approached developing subgrade models by making explicit assumptions concerning various stress, strain, and displacement variables in the theoretical equations that define the behavior of an elastic continuum. Using this conceptual approach, it is clear what the approximations and their implications are in the resulting subgrade models as shown, for example, in Table 4.1 for the three subgrade models defined to date using Reissner's conceptual approach

It is relevant to note that there is a third conceptual approach for developing simplified-continuum subgrade models and that is simply working directly with Rhines' solution (Equation 4.1). This is because simply retaining a finite number of terms from the two infinite series (not necessarily the same number on each side of the equation as was noted earlier in this chapter) that together comprise the solution to Rhines' problem is a de facto simplified-continuum model. For example, if the first two terms in each series are retained, Equation 4.1 for Rhines' solution becomes:

$$p^{II} - \frac{H^2}{3}p^{IV} = \frac{E}{H(1-v^2)}w^{II} - \frac{EH}{3(1-v^2)}w^{IV} . \tag{4.10}$$

Note that after two integrations of this equation and evaluation of the integration constants, this equation would be identical in form and order to Equation 4.3 defining the behavior of the RSC although the constant coefficients of the terms differ.

The benefit of using Rhines' solution directly to create what is hereinafter called a *Rhines-Solution Simplified Continuum* is that none of the theoretical development required for both the Reissner and Vlasov-Leont'ev approaches is required as all theoretical development was done once and for all by Rhines when solving the problem in the first place. In addition and more importantly, there is theoretically no limit as to how exact a solution can be developed although the most exact solution given explicitly in either Rhines (1965) or Kerr and Rhines (1967) retained only four terms on each side of Equation 4.1. This means that after integrating Equation 4.1 twice the most exact solution available includes derivatives of both p and w up to order four which is beyond that of the RSC which has derivatives no greater than order two (Equation 4.3).

One negative associated with using Rhines' solution directly as a conceptual approach for developing simplified-continuum subgrade models is that the explicit approximations to the elastic continuum that are clear when using either the Reissner or Vlasov-Leont'ev approach are not apparent. However, this has only academic significance and does not affect use in practice. Of more consequence is the fact that Rhines' solution is, strictly speaking, only correct for plane-strain conditions and settlements occurring in one horizontal (x) direction. On the other hand, both the Reissner and Vlasov-Leont'ev approaches yield subgrade models that are inherently correct for settlements in both the x and y directions.

Note that the issue of dealing with boundary conditions at the edge of the loaded area or foundation element (beam or plate) if one is coupled to subgrade behavior must be dealt with as with any subgrade model developed using Rhines' solution as was the case with both the Reissner and Vlasov-Leont'ev approaches.

As an aside, as noted earlier in this chapter, Rhines expressed the solution to the problem he solved in different forms in terms of how he arranged the constant coefficients used in the two equated infinite series that together comprise the problem solution. Equation 4.1 in general and the specific example shown in Equation 4.10 is but one of the forms used by Rhines and is the one most insightful for comparison to other subgrade models in this monograph. However, for reasons that will become clear in the discussion of mechanical subgrade models, it is useful to note another form of Rhines' solution that is obtained by multiplying both sides of Equation 4.1 by the term:

$$\frac{H(1-v^2)}{E} \; . \tag{4.11}$$

Using Equation 4.10 as a specific example, this equation becomes:

$$\frac{H(1-v^2)}{E}p^{II} - \frac{H^3(1-v^2)}{3E}p^{IV} = w^{II} - \frac{H^2}{3}w^{IV} \; . \tag{4.12}$$

The benefit of doing this rearrangement of the constant coefficients is that any approximate solution can then be characterized by the largest power to which the variable H (thickness of the elastic layer) is raised in the finite number of terms retained, with that power always being equal to the sum of the number of terms on both sides of the equation minus one. Thus, the lowest order possible is Order H^1 although Rhines (1965) and Kerr and Rhines (1967) call this simply Order H. Using this nomenclature, Equation 4.10, which again is of the same form as the governing equation of the RSC (Equation 4.3), in its revised form of Equation 4.12 would be termed the *Rhines-Solution Simplified Continuum of Order H^3* subgrade model.

Note that there is no upper bound of the Order to which Rhines-Solution Simplified Continuum subgrade models can be expressed. The only bounds are pragmatic and involve issues related to parameter assessment and subgrade boundary conditions that are discussed later in this chapter. As information, the highest Order model investigated by Rhines in either Rhines (1965) or Kerr and Rhines (1967) was Order H^5, i.e. three terms on each side of Equation 4.1.

4.6.5 General Considerations

4.6.5.1 Introduction and Overview

Before proceeding to a synthesis and summary of continuum-based subgrade models, there are two issues to be addressed that relevant to all such models regardless of the conceptual approach (Reissner, Vlasov-Leont'ev, Rhines) used in their development as well as important in any practical use. The first is the issue of subgrade boundary conditions at the edge of the loaded area, p. In problems where the subgrade model is coupled with the behavior of a beam or plate so that p becomes the subgrade reaction beneath the beam or plate, this boundary condition occurs at the end of the beam or edge of the plate.

The second issue involves determining values for the parameters that are used to define the coefficients in the differential equation that defines the behavior of a given model.

4.6.5.2 Subgrade Boundary Conditions

As is clear from Rhines' general solution (Equation 4.1) that provides the fundamental framework for all subgrade model behavior (not just continuum-based as discussed up to this point), any continuum-based subgrade model that retains more than one term on each side of the equation will have non-zero settlement, w, even when $p = 0$. Stated another way, there will be settlement beyond the edge of the applied load or beam or plate if there is one coupled physically and mathematically with the subgrade behavior.

In order to avoid having to explicitly model the subgrade for some distance beyond the applied load or foundation element, algebraic expressions for subgrade boundary conditions must be developed at the edge of the loaded area, beam, or plate. It is important to keep in mind that the number of boundary-condition relationships that are required increases with the increasing number of terms in the governing differential equation.

With specific regard to Reissner's approach to simplifying an elastic continuum, Reissner touched on the subject of boundary conditions briefly in his 1958 paper but discussed it in greater detail in his 1967 paper as the latter was essentially a rebuttal to two specific publications by Rhines concerning Reissner's work (Rhines 1966, 1967). In general and independent of the methodology used to develop a subgrade model, there are basically two assumptions that can be made concerning continuity of the subgrade beyond the limits of the loaded area or structural element, if any:

- vertical shear stresses within the subgrade are continuous which, depending on the specific subgrade model, can result in a discontinuity or kink in the surface settlement, w, of the subgrade at the edge of the loaded area or foundation element; and

- surface settlements are continuous which results in an upward concentrated shear force within the subgrade at the edge of the loaded area/foundation element.

The former was championed by Reissner while the latter by Rhines and his mentor, A. D. Kerr.

The writer investigated both alternatives in detail over the years (see Appendix C for some details of this) for the RSC in particular and concluded unequivocally that the former approach yields overall results that are much more in consonance with the actual behavior of foundation elements compared to the latter. Consequently, only the boundary conditions consistent with the former alternative (continuity of vertical shear stresses) are presented and utilized subsequently in this monograph.

It is unclear to the writer how continuum-based subgrade models developed using the Vlasov-Leont'ev approach fit into this broad concept for developing boundary conditions. As noted during the preceding discussion of the Vlasov-Leont'ev solution approach, two parameters that are unique to this approach, $\phi(z)$ and γ, require assumption beforehand. The roles that these two parameters play in boundary-condition development are not clear to the writer.

4.6.5.3 Parameter Assessment

A significant consideration in any application of subgrade models in either practice or research is evaluation of the constant coefficients for each term in the differential equation that defines the mathematical behavior of the subgrade model. Models that might be mathematically elegant but problematic to quantify have no pragmatic value.

The governing differential equations for subgrade models that evolve from both Reissner's approach as well as direct application of Rhines' general solution involve the traditional elastic parameters of the subgrade (E, G, ν) and the thickness of the elastic layer (or its mathematical equivalent, the depth to an assumed 'rigid' base), H. In principle, it is straightforward to evaluate each of these parameters in any specific application although doing so accurately requires some thought and effort as discussed at length in Chapter 5 and illustrated for three case histories in Chapter 6.

Parameter assessment for models developed using the Vlasov-Leont'ev approach appears to be substantially more problematic. As noted in the discussion of this methodology, there are two parameters, $\phi(z)$ and γ, that are unique to this approach and must be assumed beforehand. It is unknown to the writer if there have been any published attempts to link these unique parameters to more-conventional soil properties and problem geometry. In any event, it is clear that parameter assessment for continuum-based subgrade models based on the Vlasov-Leont'ev approach require additional, unique considerations that are beyond the scope of this monograph and thus not further discussed.

4.6.6 Synthesis

It should be obvious that the governing equations for the three subgrade models presented earlier in this chapter that are based on Reissner's conceptual approach for simplified elastic continuums...Equations 4.3 (RSC), 4.8 (PTSC), and 4.9 (F/WTSC)...each have the general form of the solution to Rhines' problem (Equation 4.1). Furthermore, each has progressively fewer terms and is thus a progressively cruder approximation to stress-displacement behavior of an actual elastic continuum.

It is also of interest that the history of developing simplified-continuum subgrade models to date has progressed from more-complex (RSC) to simpler (PTSC and F/WTSC). This is noted again as the history of developing mechanical subgrade models has been exactly the opposite, i.e. progressing from the simplest possible to the more complex as will be seen later in this chapter.

In earlier published works, the writer found that a convenient way in which to synthesize, summarize, and compare subgrade models is to display them in tabular form using the order of the derivatives of p and w that appear in the governing equation for a subgrade model. This tabular approach will be used extensively in this monograph as well.

Table 4.2 presents this tabulation for all the subgrade models identified to date by the writer that are based on any of the simplified elastic continuums that were discussed in the preceding sections. A shaded table cell indicates that the governing equation for a particular model contains a derivative of that order. Note that the derivate order for both p and w were capped at 4 in this table simply because this is the highest order model that has explicitly been identified to date in the published literature. As noted in the discussion of the Rhines Simplified Continuum, there is, in theory, no limit to the derivative orders that could be expressed using this approach.

Table 4.2. Mathematical Hierarchy of Simplified-Continuum Subgrade Models.

Subgrade Model (in groupings of mathematically identical models)	derivative order of $p(x,y)$			derivative order of $w(x,y)$		
	0	2	4	0	2	4
Fuss/Winkler-Type Simplified Continuum (F/WTSC) Rhines-Solution Simplified Continuum of Order H^1						
Pasternak-Type Simplified Continuum (PTSC) Vlasov-Leont'ev/1 layer Rhines-Solution Simplified Continuum of Order H^2						
Reissner Simplified Continuum (RSC) Vlasov-Leont'ev/2 layers Rhines-Solution Simplified Continuum of Order H^3						
Rhines-Solution Simplified Continuum of Order H^4						
Rhines-Solution Simplified Continuum of Order H^5						

4.7 MECHANICAL SUBGRADE MODELS

4.7.1 Background

As noted earlier in this chapter, mechanical subgrade models are arithmetic equations defining the aggregate stress-displacement behavior of some arbitrary assemblage of physical elements. The only limitation or requirement is that each physical element type used must have a simple, well-defined force-displacement behavior.

Note that these elements always form continuous layers that are placed in series relative to the direction of the applied stress. For consistency in the presentation of mechanical subgrade models that follows, the element layers are always labeled and discussed beginning with the layer directly beneath the applied stress (and forming the surface of the subgrade system in most cases as with mat foundations) and then progressing downward.

The elements used to date for the purpose of mechanical subgrade models fall into several categories:

- a layer of independent axial springs oriented parallel to the direction of applied loading;

- a body of liquid underlying the applied loading[33];

- a deformed[34] tensioned membrane oriented perpendicular to the direction of applied loading;

---

[33] Curiously, references that use or mention a liquid almost always refer to it in an ill-defined manner as a 'dense liquid' (no threshold value for what constitutes 'dense' is ever defined) as if there were a concern that the foundation element would 'sink' unless the liquid were sufficiently dense so that it 'floats' as with a marine vessel. But the concept of a marine vessel with a hull of finite height and thus finite freeboard does not equate with SSI applications so the issue of fluid density (actually unit weight) is irrelevant here. From a purely mathematical perspective, the relative unit weight of the fluid in SSI applications and subgrade models is unimportant and irrelevant as long as it is greater than zero.

[34] The tensioned membrane needs to be deformed in the problem formulation and concomitant derivation of the governing differential equation for a subgrade model **[continued on following page]**

- a shear-only layer oriented perpendicular to the direction of applied loading; and

- a flexure-only layer (Euler-Bernoulli or simple beam in 2-D problems and Kirchhoff-Love or thin plate in 3-D problems) oriented perpendicular to the direction of applied loading.

Note, however, that there is significant behavioral equivalency and concomitant redundancy in terms of how these element types behave arithmetically once incorporated into a subgrade model:

- a layer of springs and a body of liquid are behaviorally equivalent and thus interchangeable although the body-of-liquid visualization is rarely used nowadays and is noted here solely for the sake of completeness in view of its occasional, historical usage in the past as will be seen later in this chapter;

- a deformed/deformable tensioned membrane and shear-only layer are behaviorally equivalent and thus interchangeable; and

- the combination of a shear-only layer or deformed/deformable tensioned membrane over a spring layer is behaviorally equivalent to a flexure-only (beam or plate) layer and are thus interchangeable.

Thus, in reality, mechanical subgrade models can be created using only two different types of mechanical elements:

- a layer of springs and

- a deformed/deformable tensioned membrane.

This fact has significant practical utility for modeling mechanical subgrade models using commercially available structural analysis software as discussed in detail in Chapter 6.

Any subgrade model that contains two or more element layers is called a *multiple-parameter* or *advanced* subgrade model in the published literature. The former term is much more common as well as informative compared to the latter so is used in this monograph. As an alternative to the multiple-parameter terminology, some authors specify the number of parameters, e.g. *two-parameter model*.

Note that the term 'multiple-parameter model' is equally applicable to the simplified-continuum subgrade models discussed previously. However, the terminology tends to be more common with mechanical subgrade models simply because a much greater proportion of the published literature to date has been devoted to such models.

Historically, the term *single-parameter* model, which consists of either a layer of springs or a body of liquid, was not used for the simple reason that there was and still is only one single-parameter model and it existed for approximately 100 years before any multiple-parameter models were developed. However, for clarity in organization and presentation

---

for the same theoretical reasons of linear vs. nonlinear structural behavior discussed at length in Chapter 2 for the beam-column behavioral mechanism. Note that when such a mechanical element is implemented into a numerical solution using the matrix method then the tensioned membrane is not initially deformed in the structural model. However, it must be deformable and the deformations accounted for in a subsequent nonlinear numerical analysis as discussed in Chapter 6. The end result will then be the same as if the membrane were initially deformed.

68

within this monograph the term 'single-parameter model' will be used to unambiguously distinguish it from multiple-parameter models.

4.7.2 Organization of Presentation

As was done with subgrade models based on simplifications to a linear-elastic continuum, the presentation that follows for mechanical models is organized differently than most treatments of the subject, including the writer's earlier published works. First presented and discussed is an overview of mechanical models based on seminal research by A. D. Kerr and Rhines over 50 years ago. This is to establish the fact that an exact solution to the theory-of-elasticity problem of an applied stress on the surface of an elastic continuum that unifies all subgrade models regardless of their conceptual basis (i.e. simplified continuum or mechanical) can also be viewed as an infinite number of layers of mechanical elements. Therefore, all mechanical subgrade models, by virtue of their finite number of layers of mechanical elements, are de facto approximate solutions to this unifying problem.

This is followed by a detailed presentation of the numerous multiple-parameter models that the writer has identified to date. This presentation is organized in order of temporal evolution as, in general, researchers 'built' models of greater mechanical and concomitant mathematical complexity based on the earlier published work of others although as will be seen there were some outliers and anomalies to this otherwise logical progression of model complexity. Note that this evolutionary path is the opposite of that of simplified-continuum models, at least to date, that started with a certain level of complexity with Reissner's Simplified Continuum and then progressed temporally to simpler models (the writer's PTSC and F/WTSC).

Note also that the notation used for the various mechanical models is consistent so that comparisons between and among models is facilitated. The notation used in the original publications as well as subsequent publications by others (including the writer) is in all cases different to varying extents. The constant coefficients for the subgrade reaction, p, and settlement, w, variables are sometimes arranged differently as well in other publications.

Finally, as was done with simplified-continuum models, a summary and synthesis of multiple-parameter models is presented in tabular format.

This presentation of multiple-parameter models is followed by a presentation of the single-parameter model that is the mechanical interpretation of the Fuss-Winkler Hypothesis. This presentation is lengthier than might be expected as several different variations based on various interpretative paths have been developed over the years, all based on the same underlying model.

The treatment of mechanical models concludes with an overall summary and synthesis as well as a comparison to simplified-continuum models, both in tabular format.

4.7.3 Kerr-Rhines' Overview

Kerr and Rhines (1967) summarized the overall state of knowledge with respect to subgrade models as of that time and may well have been the first to note in an English-language publication the unifying concept for subgrade models that was discussed earlier in this chapter, i.e. that all subgrade models, whether derived using the concept of simplifying elastic continuums or resulting from arbitrary assemblages of mechanical elements, are approximate solutions to the same theory-of-elasticity problem.

Of specific relevance to mechanical subgrade models, they showed that the constant coefficients in Rhines' infinite-series solution (Equation 4.1) to the theory-of-elasticity problem that unifies all subgrade models could be defined using mechanical models. Using the Order-of-*H* concept discussed earlier in this chapter for ranking versions of the Rhines-Solution Simplified Continuum, they presented the figure that is reproduced here as Figure 4.1 (with annotations by the writer) that shows what mechanical subgrade models up to Order 10 could look like. Note that Order 10 means a total of 11 terms retained in Equation 4.1, five on the left-hand side of the equation and six on the right-hand side of the equation.

Note that Kerr and Rhines chose to primarily use combinations of spring and shear-only layers for their generalizations. The shear-only layers in particular can be replaced by deformed tensioned membranes. As discussed subsequently, there are pragmatic reasons for preferring the latter over the former in practical usage.

It is of interest to note that mechanical models of Order 5 (six terms total in Equation 4.1, three on each side) and above are not unique in terms of the assemblage of mechanical elements. This is due to the equivalent behavior of mechanical elements that was noted and discussed earlier in this chapter. In particular, a flexure-only (plate) layer can be replaced by a shear-only layer or deformed tensioned membrane that is underlain by a spring layer.

As annotated in this figure, there are additional mechanical-element substitutions that could be made beyond those noted by Kerr and Rhines so that in reality any mechanical subgrade model can be created using only two different element types, one of which must always be a spring layer or body of fluid.

4.7.4 Multiple-Parameter Models

4.7.4.1 Filonenko-Borodich

The earliest multiple-parameter model that the writer is aware of is that of Filonenko-Borodich, published in 1940. It consists of a deformed tensioned membrane over a layer of springs and has the following governing differential equation:

$$p = k_1 w - T_1 \nabla^2 w \qquad (4.13)$$

where k_1 = the spring constant and T_1 = the membrane tension[35]. This is a classic example of what some refer to in the literature as a 'two-parameter model'.

It should be readily apparent that this equation has both the general form of the unifying Rhines' theory-of-elasticity solution (Equation 4.1) and is the simplest possible multiple-parameter model, of Order H^2 using the nomenclature shown in Figure 4.1. As such, it has the form of the simplest possible subgrade model to inherently include the effects of shearing within the simulated subgrade (by virtue of the deformed tensioned membrane that links the otherwise independent 'soil springs' together) that the Fuss-Winkler Hypothesis fundamentally lacks.

It is of interest to note that Kerr and Saxena (1977)...note that this was W. C. Kerr, not the aforementioned A. D. Kerr...used the Filonenko-Borodich model for foundation-related research although they incorrectly referred to it as the Pasternak model that is discussed subsequently.

---

[35] The reason for subscripting k and T is that, in principle, it is possible to have k_i spring layers and T_i tensioned membranes in a mechanical model, where $i = \infty$ in the limit as implied in Figure 4.1.

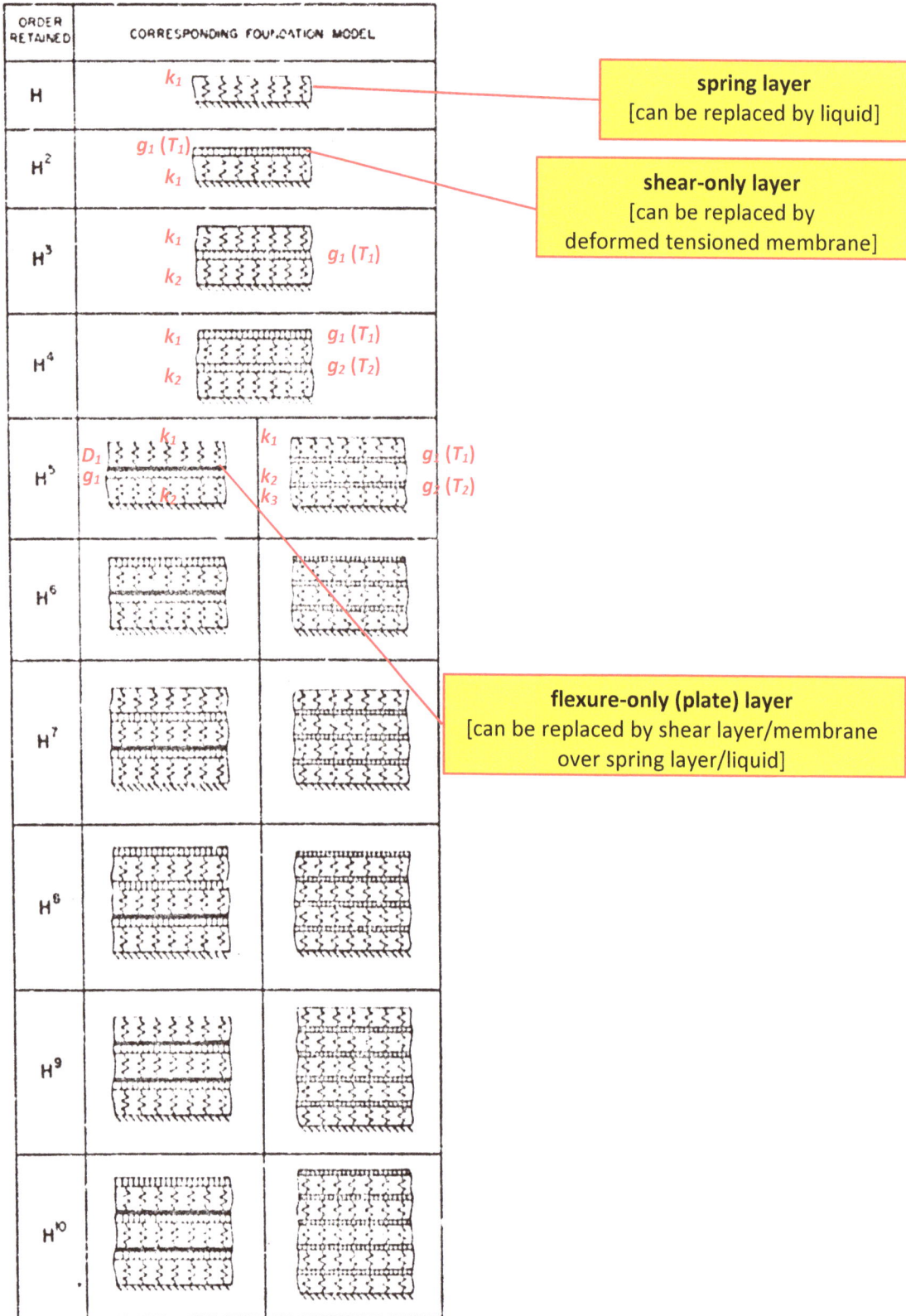

Figure 4.1. Conceptual Hierarchy of Mechanical Subgrade Models
[from Kerr and Rhines (1967)].

4.7.4.2 Hetényi

In addition to his extensive work with the Fuss-Winkler Hypothesis, Hetényi developed his own multiple-parameter model (Hetényi 1946, 1950). He postulated a model that consisted of a flexure-only plate embedded between upper and lower spring layers. The governing differential equation of this model is[36]:

$$p + \left(\frac{D_1}{k_1 + k_2}\right)\nabla^4 p = \left(\frac{k_1 k_2}{k_1 + k_2}\right)w + \left(\frac{D_1 k_1}{k_1 + k_2}\right)\nabla^4 w \qquad (4.14)$$

where D_1 = the plate stiffness (subscripted for the same reason as k_1 and T_1 for the Filonenko-Borodich model) and k_1 and k_2 = the spring stiffnesses of the upper and lower spring layers respectively.

An interesting aspect of Hetényi's model is that it is not consistent in terms of format with either Rhines' solution (Equation 4.1) or the Kerr-Rhines ranking order of mechanical models shown in Figure 4.1. This is because what would be the second (∇^2) term on each side of Equation 4.1 is missing. Therefore, the Hetényi model is an outlier in the hierarchy of mechanical subgrade models that is shown in Figure 4.1 which is why its visual depiction does not appear in that figure.

A speculative explanation for why Hetényi developed the subgrade model he did is that a good portion of his scholarly monograph on subgrade models (Hetényi 1946) was devoted to developing solutions for a flexural member (beam or plate) supported on a Fuss-Winkler subgrade which, as is well-known and will be discussed subsequently at length, is most often visualized as a bed of independent axial springs. So, a good portion of Hetényi's monograph was devoted to solutions for the problem of a plate on a spring layer. Consequently, it was logical (at least in the writer's opinion) for Hetényi to simply stack another plate + spring assemblage on top of the basic problem he was working with in order to pursue the effects of a mechanical subgrade with "continuity", as he called it, that is supporting a plate. 'Continuity' in this context is just an alternative term for 'spring coupling'.

4.7.4.3 Pasternak (A. D. Kerr Interpretation)

In 1954, P. L. Pasternak published a Russian-language paper in which he postulated the concept of a subgrade model that consisted of a layer of axial springs with shear forces acting between the springs in a direction parallel to the springs. In the context of most SSI applications such as a mat foundation, this means that the springs and inter-spring shear forces are both aligned vertically. Pasternak apparently did not suggest a physical mechanism that would produce the inter-spring forces so the writer terms this behavioral abstraction the *Pasternak Hypothesis*.

Several years later, A. D. Kerr (Kerr 1961) suggested a physical model for visualizing the abstract inter-spring shear force postulated by Pasternak. Kerr proposed an incompressible shear-only layer of unit thickness that was placed across the top of the layer of independent axial springs. The governing differential for such a subgrade model (which the writer terms the *Pasternak Model*) is:

$$p = k_1 w - g_1 \nabla^2 w \qquad (4.15)$$

---

[36] There is a typographical error in the version of Equation 4.14 that appears in Horvath (1979).

where g_1 = the shear modulus of the shear-only layer (subscripted for the same reasons as k, T, and D previously). Note that this parameter should not be confused or conflated with the shear modulus, G, of an elastic continuum. These are two completely different parameters.

It should be readily apparent that Equation 4.15:

- has the general form of the unifying Rhines' solution (Equation 4.1);

- is the simplest possible multiple-parameter model, of Order H^2 using the nomenclature shown in Figure 4.1; and

- is identical in form and thus behavior to the Filonenko-Borodich model (Equation 4.13).

This last point is not surprising as it was noted earlier in this chapter as well as in Figure 4.1 that a deformed tensioned membrane is behaviorally identical to a shear-only layer, i.e. $T_1 = g_1$.

4.7.4.4 Loof

In the early 1960s (Loof 1962, 1965), the late Prof. Henk W. Loof of the then Delft Institute of Technology[37] presented the concept of a subgrade model that is broadly identical to Pasternak's Hypothesis. In Loof's case, he postulated that the inter-spring shear forces were proportional to dw/dx, the first derivative of subgrade settlement with respect to the horizontal x axis. As with Pasternak, Loof did not offer a physical visualization of what might generate this force so the writer has termed this concept *Loof's Hypothesis*.

The writer has always considered Loof's contribution to subgrade models to be distinct from that of Pasternak despite the clear conceptual similarity and overlap and thus worthy of its own listing as a subgrade model even though the only two mechanical visualizations of Loof's Hypothesis are a deformed tensioned membrane and a shear-only layer, i.e. the Filonenko-Borodich and Pasternak models respectively. The reason for the writer's position is that Loof was more specific than Pasternak as to the source and nature of the hypothesized inter-spring shear force, i.e. that it was proportional to dw/dx. Also, Loof did not cite Pasternak's work and may well have been unaware of it, especially since Pasternak's original work was published in the Russian language and the late-1950s Cold War political situation was not conducive to interchanges of technical information between what was then the Soviet Union and Western Europe. Finally, in the pre-Internet era technology transfer moved much more slowly than it does at present. That someone could publish a Russian-language paper in 1954 and it not been known in The Netherlands by circa-1960 is not unreasonable.

In any event, there are two additional items of interest to note with regard to Loof's Hypothesis:

- He formally used the term *spring coupling* in his papers. For many years now, this term has been in common, generic, colloquial use to define any mechanism, mechanical or mathematical, for overcoming what is widely acknowledged to be the most glaring shortcoming implied by the Fuss-Winkler Hypothesis: that the cause-effect relationship between subgrade reaction and settlement at any point is independent of behavior at any other point.

---

[37] Currently known as the Delft University of Technology (TU Delft).

- Loof viewed his hypothesis as a pragmatic means to an end, not a theoretical end in and of itself as in the case of most other researchers who developed subgrade models. Loof's interest in subgrade models was related to their application in pavement engineering, specifically, the behavior of PCC pavements. He was apparently interested in developing an analytical model that was more accurate to use for rigid-pavement analysis and design for ongoing, major airfield-pavement projects in The Netherlands than the Fuss-Winkler Hypothesis and thus hypothesized the mathematical framework for a model that was one order or level more accurate than the Fuss-Winkler Hypothesis as is apparent from Figure 4.1.

4.7.4.5 Modified Pasternak/Kerr

In 1965, A. D. Kerr (1965) first suggested a new mechanical subgrade model that Rhines (1965) referred to as the *Modified Pasternak Model* (see also Kerr (1966)). In subsequent publications, including several by others (Jones and Xenophontos 1976, Gazetas 1981, Kerr 1985, Kneifati 1985), this model was referred to as the *Kerr Model* and will be so referenced in this monograph.

The Kerr Model consists of a spring layer added on top of A. D. Kerr's visualization of Pasternak's Hypothesis (shear layer over spring layer) that was discussed previously, hence the genesis of the 'modified Pasternak' terminology. The governing differential equation of the resulting spring layer + shear layer +spring layer model is:

$$p - \left(\frac{g_1}{k_1 + k_2}\right)\nabla^2 p = \left(\frac{k_1 k_2}{k_1 + k_2}\right)w - \left(\frac{g_1 k_1}{k_1 + k_2}\right)\nabla^2 w. \tag{4.16}$$

It should be readily apparent that the Kerr model is consistent with both Rhines' general solution (Equation 4.1) and the overall Kerr-Rhines hierarchy of mechanical models shown in Figure 4.1 where it is of Order H^3.

Rhines devoted much of his doctoral dissertation (Rhines 1965) to developing closed-form solutions to Equation 4.16 which is perhaps not surprising given the fact that it was a then-contemporaneous theoretical development of his mentor.

4.7.4.6 Rhines

In addition to the several contributions to the overall subject of subgrade models made by Rhines that have already been noted, he developed his own mechanical model. The *Rhines Model* as it will be called herein consists of a flexure-only plate inserted within the middle of the aforementioned Kerr Model to result in a spring layer + plate + shear layer + spring layer assemblage. The developmental details are presented in Rhines (1965) with only the final equation in Kerr and Rhines (1967).

The governing differential equation of the Rhines Model is:

$$p - \left(\frac{D_1}{g_1}\right)\nabla^2 p + \left(\frac{D_1}{k_1 + k_2}\right)\nabla^4 p \tag{4.17}$$
$$= \left(\frac{k_1 k_2}{k_1 + k_2}\right)w - \left(\frac{D_1 k_1 k_2}{g_1(k_1 + k_2)}\right)\nabla^2 w + \left(\frac{D_1 k_1}{k_1 + k_2}\right)\nabla^4 w.$$

74

Again, it should be apparent that the Rhines Model conforms to the general solution shown in Equation 4.1 and is consistent with the Kerr-Rhines hierarchy of mechanical models shown in Figure 4.1 where it is of Order H^5.

Several additional observations can be made:

- Order H^4 in the hierarchy of mechanical models is skipped over by the Rhines Model, and no model of that order has formally been developed to date although the components of that model are clearly established as shown in Figure 4.1 (note that deformed tensioned membranes could be substituted for the two shear-only layers shown in that figure if desired).

- The Rhines Model is the highest-order mechanical model formally developed to date although nothing precludes developing higher-order models as is indicated clearly in Figure 4.1.

- As shown in Figure 4.1, the Order H^5 of the Rhines Model is the lowest order for which non-uniqueness in the composition of mechanical models exists as a flexure-only plate is equivalent to the combination of a shear-only layer (or deformed tensioned membrane) over a spring layer. Note that this is in addition to and separate from the fact that a body of liquid can always be substituted for a spring layer and a deformed tensioned membrane can always be substituted for a shear-only layer as noted in this figure.

4.7.4.7 Hertz and Modified Hertz/Haber-Schaim

Haber-Schaim (1973) published a hard-cover monograph in which he proposed a subgrade model that consisted of a flexure-only plate overlying a layer of springs. This appears to be a variation of a model reportedly (Timoshenko and Woinowsky-Krieger 1959) proposed by Heinrich Hertz in 1884 who assumed the mathematically identical model of a plate 'floating' on body of liquid (hereinafter referred to as the *Hertz Model*). Therefore, Haber-Schaim's model can be considered to be an alternative form or modification of the Hertz Model.

The governing differential equation of what is herein defined as the *Modified Hertz Model* or *Haber-Schaim Model* is:

$$p = k_1 w + D_1 \nabla^4 w .$$ (4.18)

Note that the governing equation for the Hertz Model would be identical except that the variable k_1 would be replaced by γ_1 = the unit weight of the body of liquid on which the plate was 'floating'.

As with the Hetényi Model, neither the Hertz Model nor Haber-Schaim Model are consistent in terms of form with either Rhines' solution (Equation 4.1) or the Kerr-Rhines ranking order of mechanical models shown in Figure 4.1. This is because what would be the second (∇^2) term on each side of Equation 4.1 is missing. Therefore, both the Hertz Model and Haber-Schaim Model are additional outliers in the hierarchy of mechanical subgrade models that is shown in Figure 4.1 which is why visualizations of these models do not appear in that figure.

4.7.4.8 Modified Kerr/Horvath-Colasanti

The final mechanical model discussed in this monograph is one that was originally developed by the writer in 1993 although it was not publicized until almost 20 years later after additional developments rendered the model easily implemented and used in routine practice (Horvath and Colasanti 2011a). It consists of a spring layer over a deformed tensioned membrane followed by another spring layer. The governing equation of what is herein referred to as the *Horvath-Colasanti Model*[38,39] is:

$$ p - \left(\frac{T_1}{k_1 + k_2}\right)\nabla^2 p = \left(\frac{k_1 k_2}{k_1 + k_2}\right)w - \left(\frac{T_1 k_1}{k_1 + k_2}\right)\nabla^2 w \ . \tag{4.19}$$

This model is clearly mathematically identical to the Kerr Model (Equation 4.16) and thus consistent with the form of Rhines' solution (Equation 4.1) and the model hierarchy shown in Figure 4.1 (Order H^3).

Note, however, that the Horvath-Colasanti Model is not just a trivial modification of the Kerr Model (i.e. substituting a deformed tensioned membrane for a shear-only layer) as there are pragmatic reasons for preferring the use of a tensioned membrane as a mechanical element in lieu of a shear-only layer. These reasons will become clear later in this chapter. In addition, details were developed and presented in Colasanti and Horvath (2011) for handling the complex subgrade boundary conditions at the edge of a loaded area so that the subgrade beyond the edge does not have to be modeled explicitly but can be replaced by well-defined functions at the edge of the loaded area. The details of this, which are unique to the Horvath-Colasanti Model, are presented in Appendix C.

4.7.4.9 General Considerations

4.7.4.9.1 Introduction and Overview

Before proceeding to a synthesis and summary of multiple-parameter models, there are several issues that need to be addressed as they are relevant to all such models as well as important in any practical usage. The first is the issue of subgrade boundary conditions at the edge of the loaded area, p. In problems where the subgrade model is coupled with the behavior of a beam or plate so that p becomes the subgrade reaction beneath the beam or plate, this boundary condition occurs at the end of the beam or edge of the plate. This is the same issue that was discussed previously for simplified-continuum subgrade models.

The second issue involves determining values for the for the parameters (D, g, k, T) that define the mechanical elements used for the various models. Historically, this has been the biggest and most problematic issue with using mechanical subgrade models in practice.

The third and final issue discussed is time-dependent behavior of the subgrade. Although this is not something for which mechanical subgrade models have been widely used, it is a subject that needs to be addressed for the sake of completeness.

---

[38] Earlier published works by the writer called this the *Modified Kerr Model.*
[39] The writer chose to include Mr. Colasanti's name in the model name as a professional courtesy even though he was not involved in the original model development as it was his critical structural engineering contributions that made this model usable in commercially available structural analysis software as discussed in Chapter 6 and Appendix C.

4.7.4.9.2 Subgrade Boundary Conditions

As is clear from Rhines' general solution (Equation 4.1) that defines all subgrade model behavior, any mechanical subgrade model that retains more than one term on each side of the equation will have non-zero settlement, w, even when $p = 0$. Stated another way, there will be settlement beyond the edge of the applied load or beam or plate if there is one coupled with the subgrade. As discussed earlier in this chapter, this same issue arises with subgrade models developed using simplifications of an elastic continuum.

In order to avoid having to explicitly model the subgrade for some distance beyond the applied load or foundation element, algebraic expressions for subgrade boundary conditions must be developed at the edge of the loaded area or beam or plate, if any. It is important to keep in mind that the number of boundary-condition relationships that are required increases with the increasing number of terms in the governing differential equation and concomitant order of the subgrade model as reflected in Figure 4.1.

It is unclear from the published literature if boundary-condition relationships were developed for all of the multiple-parameter mechanical models discussed in the preceding sections. As developing such relationships is a prerequisite for using any model in practice, this was intentionally one of the specific goals in the development of the Horvath-Colasanti Model. The details of developing boundary-condition relationships for this model were presented in Colasanti and Horvath (2011) and can also be found in Appendix C.

As was noted previously for the RSC and other models based on simplifying elastic continuums, it should be kept in mind that the boundary-condition relationships that can be developed for mechanical subgrade models are not unique. For example, for the Horvath-Colasanti Model that is of Order H^3 per Figure 4.1 (the same order as the RSC), the choices for the edge of the loaded area are the same as they were for the RSC:

- continuity of surface settlement, w, or

- continuity of vertical shear forces.

In general, the differences in results between the two assumptions are significant and should always be investigated thoroughly before selecting one to use. As noted previously, the writer has found that the latter assumption generally yields calculated results that are more in consonance with actual SSI behavior than the former assumption.

4.7.4.9.3 Parameter Assessment

Quantifying the constant coefficients that appear in the governing differential equations of multiple-parameter mechanical subgrade models is inherently problematic and ill-defined as the correlation between the parameters (D, g, k, T) that define the behavior of each mechanical element and conventional soil properties are not obvious. As a result, the issue of parameter assessment for multiple-parameter mechanical models has long been the proverbial 'elephant in the room' that is simply ignored in some cases (some researchers do not even discuss what the subgrade being modeled might consist of, no less mention soil in particular) and discussed in a relatively superficial way of limited practicality in other cases.

In general, when the issue of parameter assessment has been addressed the overall approach taken is to rely on the mathematical process of *collocation* that involves developing a best-fit match of model parameters to actual stress-displacement behavior measured in some field test. More often than not, this field test that is recommended or proposed is the traditional plate load test (Kerr 1985). Rhines (1965) claims that this approach was suggested by Pasternak for use with what subsequently evolved as Kerr's interpretation of the Pasternak Hypothesis (Equation 4.15) and pursued this approach to a limited extent in his own doctoral dissertation. The work of both Loof (essentially Equation 4.15) and Haber-Schaim (Equation 4.18) is notable as they explicitly discussed parameter assessment for their models in some detail, in each case focusing on the plate load test exclusively.

The overall problem with using the concept of collocation in general for quantifying subgrade model parameters is that for the final outcome to be accurate the test parameters in terms of the dimensions of the loaded area, depth of embedment, stress level applied, etc. must mimic those of the intended foundation element. If not, then there are well-known scaling issues related to a mismatch between test and intended application in terms of stressed depth of the subgrade, stress levels, and other significant factors. The specific issue with using the concept of collocation based on plate load test data is that such tests are traditionally performed using a plate or stack of plates with a maximum diameter of 30 inches (762 mm). As such, this test is intended primarily for pavement design. The practical impossibility of scaling-up results from such tests for applications such as mat foundations have been well and widely known in foundation engineering for decades.

Consequently, the use of the collocation concept based on plate load test data is only a meaningful approach when the scale of the actual loading is of the same order of magnitude as the aforementioned plate dimensions. One such application is that of airfield pavements which was the sole application of interest to Loof (1962, 1965). Consequently, Loof was able to demonstrate good correlation between measured and forecast results using his subgrade hypothesis which, as noted previously, is essentially the same as Kerr's visualization of Pasternak's Hypothesis. On the other hand, collocation based on plate load test data for applications such as mat foundations is essentially worthless.

In conclusion, in the writer's opinion the issue of parameter assessment is, as it has always been, a significant problem with any and all mechanical models. The problems are only amplified as the Order of the model (as reflected in Figure 4.1) increases and the number of parameters to be evaluated increases accordingly.

4.7.4.9.4 Viscoelastic Behavior

As noted in the discussion of subgrade models based on a simplified elastic continuum, it is possible to incorporate time-dependent (viscoelastic) subgrade behavior into subgrade models. In fact, this was an integral component of Reissner's original (1958) paper on the subject although Reissner did not pursue solutions and application to the viscoelastic problem using the RSC.

Viscoelastic behavior has been investigated with multiple-parameter mechanical subgrade models to a modest extent. It appears that A. D. Kerr was an early researcher of this extension of mechanical models, specifically using Pasternak's Hypothesis (Kerr 1961, 1964). Furthermore, it is no surprise that Kerr's protégé, Rhines, explored viscoelastic behavior of the Kerr Model in his doctoral dissertation (Rhines 1965).

It is noteworthy that both A. D. Kerr and Rhines pursued the implementation of viscoelastic behavior to a much greater extent that Reissner did. Not only were various time-dependent relationships pursued, there was some comparison to loading of full-size foundation elements under condition where time-dependent behavior could be expected. It appears that this research was motivated by the needs and interests of the funding agencies for their work during the 1960s timeframe so included applications such as shallow foundations supported on non-earth subgrades such as snow.

4.7.4.10 Synthesis

Table 4.3 summarizes the mechanical elements used for the multiple-parameter mechanical subgrade models presented in the preceding sections. The Order shown is that proposed by Kerr and Rhines as shown in Figure 4.1 Note that the Filonenko-Borodich Model and the hypotheses of Pasternak and Loof are mathematically identical as are the models of Kerr and Horvath-Colasanti as well as Hertz and Haber-Schaim. Note also that the models of Hertz, Haber-Schaim, and Hetényi are outliers in a logical progression of the Order and concomitant complexity of subgrade models as was noted previously.

Table 4.3. Mechanical Elements for Multiple-Parameter Mechanical Subgrade Models.

Order	Subgrade model (in groupings of mathematically identical models)	Mechanical elements used to visualize model (as viewed from top to bottom)
H^2	Filonenko-Borodich	deformed tensioned membrane + springs
	Pasternak's Hypothesis / Loof's Hypothesis	shear layer + springs
H^3	Kerr (aka Modified Pasternak)	springs + shear layer + springs
	Horvath-Colasanti (aka Modified Kerr)	springs + deformed tensioned membrane + springs
-	Hertz	plate + liquid
	Haber-Schaim (aka Modified Hertz)	plate + springs
-	Hetényi	springs + plate + springs
H^5	Rhines	springs + plate + shear layer + springs

Table 4.4 ranks the multiple-parameter mechanical subgrade models in terms of the order of the derivatives of p and w as was done previously for the continuum-based models in Table 4.2.

Table 4.4. Mathematical Hierarchy of Multiple-Parameter Mechanical Subgrade Models.

Order	Subgrade Model (in groupings of mathematically identical models)	derivative order of $p(x,y)$			derivative order of $w(x,y)$		
		0	2	4	0	2	4
H^2	Filonenko-Borodich / Pasternak's Hypothesis / Loof's Hypothesis						
H^3	Kerr (aka Modified Pasternak) / Horvath-Colasanti (aka Modified Kerr)						
-	Hertz / Haber-Schaim (aka Modified Hertz)						
-	Hetényi						
H^5	Rhines						

The 'building block' approach to mechanical subgrade model creation that was shown visually in Figure 4.1 and verbally in Table 4.3 is apparent mathematically in Table 4.4. Note, however, that because the Hertz, Haber-Schaim, and Hetényi models omit some intermediate terms these models are not necessarily more accurate than models of lower Order. This is because they are missing some of the lower-Order behavioral characteristics.

4.7.5 Single-Parameter (Fuss-Winkler) Model

4.7.5.1 Governing Equation and Physical Interpretations

It is apparent from the preceding discussion of multiple-parameter mechanical models and the visual summary presented in Figure 4.1 that there is only one possible single-parameter mechanical model, i.e. a model that retains only one term on either side of Rhines' general solution (Equation 4.1). This single-parameter mechanical model is of Order H^1 (H using the original Kerr-Rhines nomenclature shown in Figure 4.1) and is thus the simplest mechanical subgrade model possible.

The governing differential (technically) equation of this single-parameter mechanical model is:

$$p = k_1 w \qquad (4.20)$$

which is clearly just a mechanical visualization of the abstract Fuss-Winkler Hypothesis (Equation 4.2) that was discussed at length earlier in this chapter so will be referred to as the *Fuss-Winkler Model*. Thus, the k_1 parameter in Equation 4.20 is the conceptual equivalent of the k_{FW} parameter in Equation 4.2 but with some significant differences in the implications and interpretations of k_1 vs. k_{FW} as discussed subsequently.

Note that Equation 4.20 expresses the near-universal visualization of the Fuss-Winkler Model as a layer of independent axial springs. An alternative mechanical visualization of the support medium is a body of liquid (as used in the Hertz multiple-parameter model noted earlier) in which case k_1 is replaced by the liquid unit weight (more commonly referred to as 'density' in the published literature which is technically incorrect on a dimensional basis), γ_1.

To the best of the writer's knowledge, it is unclear who first visualized the abstract Fuss-Winkler Hypothesis as either a spring layer or body of liquid and thus 'created' the Fuss-Winkler Model. It is clear from Hertz's work that the body-of-liquid analogy at least existed by late in the 19th century.

It is also clear that the spring-layer analogy has long been the predominant and preferred visualization and mechanical interpretation of the Fuss-Winkler Model. This has given rise to numerous colloquial terms for the k_1 parameter in Equation 4.20, usually 'soil spring constant' or some variation of this. Unfortunately, this has created the widespread perception that this parameter is somehow an inherent soil property. In reality, nothing could be farther from the truth. Nevertheless, this perception endures and is thus one of the longest lived and most widespread 'unicorns' of geotechnical and foundation engineering.

4.7.5.2 Traditional Assumptions, Usage, and Parameter Assessment

Historically and traditionally (e.g. ACI Committee 336 1988), the k_1 parameter in Fuss-Winkler Model is assumed to be spatially uniform and load-wise constant in a given problem application. However, as noted in the prior discussion of the Fuss-Winkler

Hypothesis there is nothing inherent in either the original hypothesis (Equation 4.2) or the mechanical visualization of same (Equation 4.20) that requires either assumption. These were simply pragmatic, even necessary, assumptions in the pre-computer era to allow development of closed-form solutions. However, as Hetényi showed in his monograph on subgrade modeling and models (Hetényi 1946), it is possible to develop closed-form solutions for the Fuss-Winkler Model assuming a simple spatial (but not load-wise) variation in k_1.

Given the fact that the k_1 parameter is typically assumed to be an application-specific constant, there has been decades of research and countless publications of all types devoted to the subject of evaluating this parameter. The logic applied to this process has typically been based on the assumptions that this parameter is:

- an inherent subgrade (soil in most cases) property or

- can be correlated with an inherent subgrade/soil property or

- can be correlated with the measured results of some common geotechnical site-characterization tool such as the Standard Penetration Test N-Value.

Some of the writer's earliest publications on subgrade models (e.g. Horvath 1988c) discuss and compare some of the published work by others in this regard.

In simple terms, this collective effort over many decades has been an exercise in futility for at least two fundamental reasons:

- the fact discussed at length in Chapter 3 that the subgrade reaction, p, in a given application is a factor of the flexural stiffness of the foundation element and not just the stiffness of the subgrade alone and

- the inherent, fatal deficiencies in the traditional assumption of k_1 = constant as noted earlier in this chapter in the discussion of the Fuss-Winkler Hypothesis.

The following sections discuss the supporting logic behind these statements and discuss the analytical strategies that have been developed in recent decades in an attempt to overcome these deficiencies. As will be seen, these strategies have met with widely varying success that depends largely on the specific SSI application.

4.7.5.3 Model Deficiencies

The genesis and crux of the problems surrounding the traditional use of the Fuss-Winkler Model is the basic fact that although subgrade models are not complete constitutive (soil) models, they need to at least adequately replicate the essential aspects of subgrade behavior that are most relevant to a particular SSI application. For the mat foundation example used throughout this monograph, this means adequately replicating the physical mechanisms of resisting 1-D (vertical) compression as well as load redistribution in horizontal directions...what is colloquially called *load spreading*...by vertical shear stresses that develop within the subgrade.

For subgrade models based on an elastic continuum, Reissner's approach in particular makes assumptions concerning this crucial vertical-shearing capability completely obvious. As can be seen in Table 4.1, both the original RSC Model and as well as the PTSC

Model developed by the writer include this crucial behavior. On the other hand, the F/WTSC Model developed by the writer clearly does not.

With mechanical models, the presence or absence of the vertical-shearing/load-spreading mechanism is reflected in the presence or absence of what is colloquially referred to (since at least circa 1960) as what Loof called 'spring coupling'. As can be seen in Figure 4.1, all mechanical models have at least one layer of independent axial springs that replicates the required 1-D subgrade-compression mechanism. If there is another type of mechanical element (deformed tensioned membrane, shear-only layer, flexure-only layer) that connects the tops of the otherwise independent springs together, this replicates the mechanism of spring coupling. Furthermore, the more layers of these connective mechanical elements that exist the better the approximation of the load-spreading/vertical-shearing behavior within an actual subgrade. This is because, in the limit, an actual subgrade consists of an infinite number of these connective layers as can be seen in Figure 4.1.

It is apparent in Figure 4.1 that all multiple-parameter mechanical models inherently include at least a basic or primitive amount of spring coupling in their physical visualization and, therefore, derived mathematical behavior. On the other hand, the single-parameter Fuss-Winkler Model does not. This, then, is the most significant fatal flaw in this single-parameter mechanical model (and the abstract Fuss-Winkler Hypothesis by extension), the lack of inherent spring coupling.

The essential role that vertical shearing plays in actual SSI applications...and, by extension, the essential role played by inherent spring coupling in multiple-parameter subgrade models (both mechanical and simplified continuum)...can be seen in the simple limiting cases shown in Figure 3.4 that is replicated here as Figure 4.2 for ease of reference.

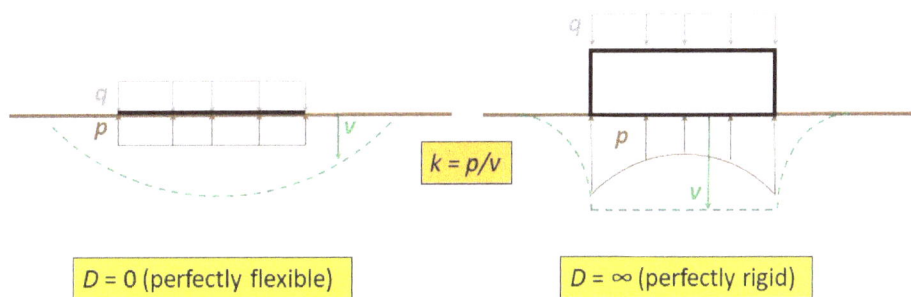

$$k = p/v$$

$D = 0$ (perfectly flexible) $D = \infty$ (perfectly rigid)

Figure 4.2. Qualitative Aspects of Actual Subgrade Reaction and Settlement.

It is obvious that whether the external load, q, is applied to either a perfectly flexible or perfectly rigid foundation element (and, by implication, any intermediate value of plate stiffness $0 < D < \infty$) that the generic coefficient of subgrade reaction, k, that is the ratio of subgrade reaction, p, to surface settlement, w, is never spatially uniform in magnitude. Furthermore, it is obvious that the subgrade reaction, p, does not depend on the subgrade behavior alone but is also a reflection of the relative stiffness (D in this case) of the foundation element in contact with the subgrade which is the essence of SSI. Therefore, it is reasonable to expect that a technically adequate subgrade model should replicate at least the basic elements of the type of behavior reflected in Figure 4.2.

The writer's research dating back to the publication of Horvath (1979) indicates that multiple-parameter models, whether based on a simplified elastic continuum or an assemblage of mechanical elements, are inherently capable of replicating the effects of subgrade shear. Furthermore, it is not surprising that the more parameters in a model, the better it is at replicating subgrade shear and thus the generic behaviors shown in Figure 4.2. This is to be expected as the more parameters in a subgrade model the better approximation it is to the infinite-series exact solution developed by Rhines (Equation 4.1). On the other hand, the single-parameter mechanical model with a constant value of k_1 in Equation 4.20 as assumed historically and traditionally will never be able to inherently replicate the behaviors shown in Figure 4.2.

Perhaps the most extensive theoretical work that illustrates the inability of the Fuss-Winkler Model with k_1 = constant to replicate actual subgrade behavior was the extensive work by the late Prof. Aleksandar Vesic and co-workers. Vesic's relatively obscure published work related to 3-D applications involving plate-type foundation elements (Vesic and Saxena 1970[40]) determined that, in general, three separate values of k_1 were required in a given problem to match bending moment, settlement, and subgrade reaction between the Fuss-Winkler Model and an isotropic, homogeneous elastic half-space. The range in magnitude between the largest and smallest k_1 values was of the order of six. Furthermore, this match was achieved only directly beneath the applied load as the match was poor elsewhere along the plate-subgrade interface.

This body of theoretical work is complemented and supported by the observation of the behavior of actual foundation elements in SSI applications (Vesic and Johnson 1963; Horvath 1988b, 1988c, 1993c, 1993d; Liao 1995). Collectively, both theory and reality explain and support the statement made previously that the decades of effort to find such a single-valued 'answer' for k_1 = constant in the Fuss-Winkler Model have been inherently exercises in futility that is akin to ever finding the mythical unicorn.

For the sake of completeness, it should be noted that separate from the inherent futility of uniquely linking the k_1 parameter to fundamental soil properties or site-characterization measurements, there is the additional issue that published correlations have often been misinterpreted or otherwise misused in practice. One such instance with regard to Terzaghi's published correlations for k_1 was noted in the discussion of the Fuss-Winkler Hypothesis earlier in this chapter. Terzaghi clearly intended for the Fuss-Winkler Model and his recommendations to be used only for a very narrow structural analysis purpose of calculating local bending moments yet his correlations for k_1 have been routinely published in textbooks and design manuals as all-purpose recommendations in SSI applications.

Another notable example of chronic misinterpretation and misrepresentation involves the aforementioned published work of Vesic. The relative obscurity and concomitant general lack of knowledge of Vesic's published work related to the 3-D problem of plate-type foundation elements such as mats and rigid pavements (Vesic and Saxena 1970) has created some confusion in the subsequent published literature by others. Numerous researchers and textbook authors have mistakenly used Vesic's much better-known work on the 2-D problem involving beam-type foundation elements (Vesic and Johnson 1963), which involves an empirical 12th-root equation, for plate-type foundations which involves an empirical cube-root equation. There are significant differences between these two Vesic solutions as discussed in detail in Horvath (1988b, 1988c). This confusion has been all the more unfortunate because plate-type shallow foundations such as mats, slabs-on-grade, and rigid pavements are much more common in practice compared to beam-type shallow foundations such as combined footings.

---

[40] A more accessible abstract can be found in Scott (1981, pp. 161-162).

4.7.5.4 Implementation in Structural Analysis

4.7.5.4.1 Background and Overview

One of the most significant reasons for the enduring popularity of the Fuss-Winkler Model among practitioners and academicians alike despite its inability to inherently replicate actual subgrade behavior is the ease with which it can be implemented into both simple and state-of-art structural analyses. This point is nicely summed up by the following statement in Brown (1974):

"For foundation problems, the Winkler assumption remains what it has always been, one of mathematical convenience rather than physical reality."

In the writer's opinion, this is a large part of the reason why significant intellectual energy and effort, especially by technical committees of professional organizations such as the American Concrete Institute (ACI) that represent several stakeholders in SSI applications, have been expended to develop work-around analytical strategies for the Fuss-Winkler Model such as discussed in the following sections when logic would dictate that this model should have been abandoned decades ago. Therefore, before proceeding with further discussion related to the Fuss-Winkler Model a discussion of the basic ways in which this model is implemented in structural analyses is of value. For simplicity and without any limit on generality, the 2-D problem of a beam will be used as an example. The extension to the 3-D problem involving a plate is straightforward.

4.7.5.4.2 Details of 2-D Implementation

To begin with, reference is made to Figure 2.4 that shows the general problem of a beam-column with subgrade support. If the axial force, P, = 0, this becomes the special case of just a beam of uniform flexural stiffness, EI, with only transverse loading and subgrade support. The governing differential equation of this problem is a simplification of Equation 2.12:

$$EI \frac{d^4w}{dx^4} + p = q \,. \tag{4.21}$$

Note that w, p, and q are all functions of x but this is not shown explicitly in keeping with the simplified notation adopted in this chapter for clarity.

Substituting Equation 4.20 for the Fuss-Winkler Model into Equation 4.21 to eliminate p as a variable yields:

$$EI \frac{d^4w}{dx^4} + k_1 w = q \tag{4.22}$$

that represents the coupled behavior of a beam on a Fuss-Winkler subgrade. Note that the subgrade contribution has been completely eliminated from explicit consideration and replaced by a constant coefficient, k_1, times the settlement variable, w.

Given the extensive use of computer software in structural analysis and design nowadays, it is even more insightful to view this alternatively using the matrix-method formulation that was discussed in Chapter 2. Using matrix notation, Equation 4.21 becomes:

$$[S]\{d\} + \{p\} = \{q\}. \tag{4.23}$$

Here, the subgrade-reaction vector, $\{p\}$, is defined using Equation 4.20 for the Fuss-Winkler Model that in matrix notation becomes:

$$\{p\} = [k_1]\{d\}. \tag{4.24}$$

Substituting Equation 4.24 into Equation 4.23 yields:

$$[S]\{d\} + [k_1]\{d\} = \{q\} \tag{4.25}$$

that simplifies to:

$$[S']\{d\} = \{q\} \tag{4.26}$$

where $[S']$ is the modified or enhanced stiffness matrix and is defined as follows:

$$[S'] = [S + k_1]. \tag{4.27}$$

Thus, all the subgrade effects are easily incorporated into the stiffness matrix of the foundation element (beam in this case). Although the details are not shown here, this entails simply adding k_1 to various elements in the original stiffness matrix that was discussed in Chapter 2.

The modifications to the element of the basic stiffness matrix, $[S]$, that was defined in Chapter 2 (Equation 2.4 for an Euler-Bernoulli or simple beam and Equation 2.7 for a Timoshenko beam) that are reflected in Equation 4.27 can be visualized as adding an external, independent spring at each node of the foundation element (ACI Committee 336 1988, 1989; Bowles 1988). The orientation of each spring is perpendicular to the foundation element-subgrade interface. Commercially available structural analysis software can easily accommodate such springs.

There are two comments to close out this discussion that provide a convenient segue to items discussed in the following sections:

- In the basic case defined by Equation 4.27, each spring has the same, constant magnitude, k_1. However, in an actual application using the matrix method of analysis it is easy to specify different values of k_1 at different nodes along the foundation-subgrade interface.

- The concepts presented above could easily be extended to a beam-column (the $P \neq 0$ case shown in Figure 2.4). Equations 4.23 through 4.27 also apply to a beam-column but in this case the basic stiffness matrix, $[S]$, is the one defined by Equation 2.9.

4.7.5.5 Analytical Strategies to Compensate for Fuss-Winkler Model Deficiencies

4.7.5.5.1 The Pseudo-Coupled Concept: Basic Approach

It is clear from the preceding discussion that the most glaring and significant shortcoming of the Fuss-Winkler Model is its inability to inherently replicate shearing resistance and concomitant load spreading that is arguably the single most important behavioral mechanism of actual subgrades in SSI applications. As noted previously, this shearing resistance is colloquially referred to in mechanical subgrade modeling as spring coupling and, thus, it is typically stated that the Fuss-Winkler Model inherently lacks spring coupling in its mathematical formulation.

In the latter decades of the 20th century, an analytical concept evolved that is referred to as *pseudo-coupling*. In simple terms, the *Pseudo-Coupled Concept*, as it will be referred to in this monograph, seeks to retain the structural-analytical simplicity of the Fuss-Winkler Model that was illustrated in the preceding section but to compensate for its lack of inherent spring coupling by:

- requiring spatially varying values of the k_1 parameter in Equation 4.20 in order to mimic the effects of spring coupling and thus produce calculated values of subgrade reaction, p, that better match observed results and

- in the process, placing the entire burden of what this spatial variation of k_1 should be on the design professional or academic researcher performing the analysis.

It is not facetious to say that the Pseudo-Coupled Concept essentially requires knowing the correct answer beforehand in order to come up with the correct values of k_1 that will produce the correct answer. This is because both the relative spatial variation as well as absolute magnitudes of k_1 must be correctly estimated beforehand. Sometimes this is expressed as determining the project-specific absolute magnitude of some base or reference value of k_1 and then applying to this base value some relative, dimensionless spatial-variation factor across the entire foundation-subgrade interface.

Putting aside the obvious demands and challenges imposed by use of the Pseudo-Coupled Concept, it is relevant to note that fundamentally at least the concept is nothing new. As noted earlier in this chapter, there is nothing inherent in the original Fuss-Winkler Hypothesis or the Fuss-Winkler Model that derives from it that requires the k_1 parameter to be spatially constant in a given problem. However, what is new is that the Pseudo-Coupled Concept is focused on developing spatial variations in the k_1 parameter specifically to mimic the missing spring-coupling effects.

There is no doubt that the Pseudo-Coupled Concept can, in principle, produce reasonable results. As was discussed in Chapter 3 and illustrated by Equation 3.1, the coefficient of subgrade reaction, k, when viewed in its true, generic form as a calculated outcome of an SSI analysis is a spatially varying parameter. As will be seen in the following sections, the practical issue is whether or not it is realistic to create a database of correct answers based on field measurements and/or more-advanced analyses for a given category or type of SSI application so that reasonably correct values for the k_1 parameter when used in the Fuss-Winkler Model as an input parameter can be estimated in a reasonably simple manner for use in future uses of that SSI application. This is because experience has indicated that SSI applications are so diverse that it is not possible to create a one-size-fits-all approach to the problem although, as will be seen, this has been attempted, at least to some extent.

4.7.5.5.2 The Pseudo-Coupled Concept: Generic Application

As best as can be determined, the primary developers and promoters of the Pseudo-Coupled Concept, at least in U.S. practice and for generic shallow-foundation applications but especially mats, were ACI Committee 336 - Footings, Mats, and Drilled Piers as a group and some of its then-prominent members individually. The earliest known publications promoting the concept for use with mats were in the final decades of the 20th century (ACI Committee 336 1988, 1989; Bowles 1988).

It appears that the logic used to promote this generic use of the Pseudo-Coupled Concept is that although each foundation application is unique, it is (assumed) to be possible to develop generic patterns or distributions for the k_1 parameter, at least for relatively simple foundation geometries and superstructure load distributions. It is then argued that these results are of sufficient accuracy for general use in practice. Typically, these patterns in subgrade reaction are represented as some dimensionless variation in k_1 relative to a problem-specific base or minimum value of this parameter. Of course, this still leaves the problem of how to rationally determine what this base value should be.

The generic variations suggested in the late 1980s were all developed with mat foundations in mind and assumed an increase in k_1 near the edges of the mat. This derives from the fact that, for relatively simple loading, the observed subgrade reaction, p, tends to increase toward the edges of a shallow-foundation element, assuming that it has some nominal embedment below the ground surface. Note that this observed trend is also consistent with generic, theoretical results as shown in Figure 4.2 for the perfectly rigid case.

The simplest suggestion made is to merely double the otherwise spatially uniform base value of k_1 along the edges of the mat (ACI Committee 336 1988; Bowles 1988). The arbitrary and ambiguous nature of this suggestion was noted by Liao (1995).

A somewhat more sophisticated approach is to use a variation in k_1 that is based on the theory of linear elasticity. Usually the solution for a perfectly flexible, uniformly loaded area on an isotropic, homogeneous layer of finite thickness is used. Typically, this produces values of k_1 that are also about twice as large around the edges of a mat compared to its base value at the center but with a gradual, curved pattern of variation in between. Again, note that this curved pattern is consistent with theoretical results shown in Figure 4.2. Such an approach is also discussed in ACI Committee 336 (1988, 1989) and Bowles (1988).

Regardless of the actual assumption used, there is still the problem of how to rationally determine a base value for k_1 at the center of the foundation element. The various methodologies that were available during the timeframe when the Pseudo-Coupled Concept was being promoted and debated are discussed in some detail in Horvath (1979, 1988b).

The conclusion reached by the writer many years ago was that mat foundations do not lend themselves to a generic application of the Pseudo-Coupled Concept along the lines developed and promoted by ACI Committee 336 in the late 1980s. While there are various methodologies that might be used to estimate an average baseline value of the k_1 parameter on a project-specific basis, the geometry and loading of mats tends to be too unique and individualistic to lend itself to using the simplistic relative distributions of the k_1 parameter as recommended by ACI Committee 336 and Bowles. Furthermore, this individualistic nature of mat foundations discourages developing a database that could be used to forecast the spatial variation of k_1 reliably on a project-specific basis as the potential range in spatial variations is just too large.

4.7.5.5.3 The Pseudo-Coupled Concept: Specialized Applications

Although the Pseudo-Coupled Concept in its generic form as proposed by ACI Committee 336 and applied to mat foundations did not turn out to be the analytical improvement for design professionals that it was presumably intended to be, experience has shown that the concept can be useful in both routine practice and academic research for certain SSI applications in certain circumstances. This section discusses three specific situations where the Pseudo-Coupled Concept either was or still is successful in providing a workable analytical methodology for 'fooling' the Fuss-Winkler Model into exhibiting the spring-coupling behavior it does not inherently possess.

The first example involves the base slabs of 'box' tunnels with a rectangular cross-section constructed using the traditional cut-and-cover method. In this SSI application, a base slab is effectively a mat foundation with relatively simple geometry, loading, and structural continuity with the side walls and interior wall(s), if any, of the overall tunnel structure. One of the additional behavioral and concomitant analytical simplifications of box tunnels is that because of their relatively large length-to-width ratio combined with their relative rigidity in the longitudinal direction, they are typically designed using a 2-D plane-strain analysis of a typical transverse section through the tunnel. This means that base-slab flexure occurs only in one (the transverse) horizontal direction which reduces the behavioral demands on a subgrade model in this application.

In the 1990s, the Central Artery/Third Harbor Tunnel (CA/T) Project in Boston, Massachusetts (referred to colloquially as the Boston Big Dig) required the design of substantial numbers and lengths of box tunnels. Given the complex organizational structure of the CA/T Project with design, peer review, and oversight spread over many different organizations, there was a desire by the overall project management team to develop a single, project-wide analytical methodology for tunnel base slabs that optimized the usual conflict between accuracy and ease of use of SSI analytical methodologies that design professionals face on every project.

One of the lead design professionals involved in this effort was Dr. Samson S. C. 'Sam' Liao, P.E., a protégé of the writer[41] who was directly involved with the CA/T Project. He was also the Principal Investigator for an in-house research study of tunnel base slabs (Liao 1991, 1995) that found that for slabs that are stiff relative to the subgrade, the type of loading (uniformly distributed vs. concentrated loads) was not an important factor.

Consequently, Dr. Liao developed a series of chart and tabular solutions for a uniformly loaded, infinitely long strip resting on a homogeneous, isotropic, linear-elastic continuum of finite thickness. This continuum problem was solved using the FEM. These FE solutions provide relative values of the variation of subgrade reaction, p, and thus the coefficient of subgrade reaction for the Fuss-Winkler Model, k_1, beneath the simulated tunnel base slab. Specific sections of tunnel base slabs were then designed using commercially available structural analysis software that incorporated the Fuss-Winkler Model with spatially varying k_1 values as the tunnel base slab subgrade model.

With some judgment and simplification, Liao's results provided a spatially uniform value for k_1 within the middle 60% of the slab, with increasing values between there and the edge of the base slab. However, as with the generic ACI Committee 336/Bowles distributions discussed previously, there was still the problem of how to determine the reference value of k_1 at the center of the base slab. Liao offered his own suggestions on the topic.

As a final comment about this CA/T Project application, it is of interest to note that there was not universal agreement with using the Fuss-Winkler subgrade model in general

---

[41] In the interest of full disclosure and transparency, circa 1990 the writer had several meetings and other conversations and communications with Dr, Liao about his work with tunnel base slabs on the CA/T Project. This contact was initiated by Dr. Liao as a result of his prior professional working relationship with the writer and knowledge of the writer's background and expertise related to subgrade modeling and models. However, the writer was never an official, paid consultant for this aspect of work on the CA/T Project. The CA/T Project paid consulting work performed by the writer approximately a decade later (circa 2000) involved an entirely different aspect of design work related to the use of block-molded expanded polystyrene (EPS-block) as a lightweight-fill (geofoam) material for earthwork construction on above-ground portions of roadway.

and the outcomes of Liao's work in particular for project design. There was a not-insignificant school of thought at the time among some influential project stakeholders that the outdated Conventional Method of Static Equilibrium that was mentioned earlier in this chapter and is discussed in some detail in Appendix B be used. The argument presented at the time for this was the hoary chestnut of 'because it's more conservative' that is debunked in Appendix B.

The second specific application of the Pseudo-Coupled Concept noted here is the Discrete Area Method that was discussed in some detail in Chapter 3 as part of the Pseudo-Ideal analysis alternative. Although the Discrete Area Method is its own analytical methodology, it can also be viewed as a Pseudo-Coupled Concept strategy so is mentioned here for the sake of completeness.

As discussed in Chapter 3, the only known use to date of the Discrete Area Method has been with mat foundations. It addresses the individualistic nature of mats that doomed the overly simplistic, generic approaches recommended by ACI Committee 336 by requiring a project-specific determination of both the magnitude and variation of the k_1 parameter used by the structural engineers in their analyses through an iterative, trial-and-error process that requires a parallel analysis by geotechnical engineers. Because the geotechnical engineers can, in principle, use state-of-art geotechnical software that analyses a complete 3-D subgrade, it is possible to obtain very accurate results.

In summary, the Discrete Area Method can be viewed as the best possible outcome of the Pseudo-Coupled Concept as accurate, project-specific magnitudes of the k_1 parameter in the Fuss-Winkler Model that mimic the effects of spring coupling to a high degree of accuracy are achievable. However, this comes at a relatively steep cost of time and effort.

The third and final special application of the Pseudo-Coupled Concept is what is arguably the best development of this concept to date and indicates what can be accomplished using this concept if an adequate database of 'correct' answers is able to be created. This is the well-known *p-y Curve* method for the analysis of laterally loaded deep-foundation elements.

The application of the Fuss-Winkler Model to laterally loaded deep foundations is an example of a type of SSI application where a subgrade model is used to remove not one but two spatial variables from consideration in what is inherently a 3-D problem in reality. This is because the subgrade model removes both horizontal directions (parallel and transverse to the direction of loading) from explicit consideration and lumps all of this behavior into subgrade 'springs' that exist only in the depth-wise (z-axis) direction. Furthermore, these depth-wise springs lack inherent spring coupling.

The reason why the Pseudo-Coupled Concept has been so successful with laterally loaded deep foundations is that despite the wide variety of deep-foundation types and sizes, ground conditions, etc., it has been possible to amass a database using the results from countless instrumented load tests performed in a wide spectrum of terrestrial and marine conditions. Furthermore, development of this database dates back to at least the 1950s as the behavior of laterally loaded deep foundations (limited to driven piles initially) was essential to the development of the offshore oil production industry in the U.S. and thus well-funded. In recent decades, the need to understand the lateral-load behavior of all types of deep-foundation elements for roadway applications and for all types of structures under seismic loading has continued and broadened the financial support for testing of drilled shafts and other drilled types of deep foundations, especially by the Federal Highway Administration and state departments of transportation. As a result of these many decades of research, it has literally been possible to create a massive database that allows what is necessary in principle to successfully use the Pseudo-Coupled Concept with the Fuss-Winkler Model, i.e. to know the correct answer beforehand so that the correct input parameters can be chosen to produce the correct answer.

4.7.5.5.4 The Beam-Column Analogy

As noted previously, one of the primary reasons for the enduring attraction of the Fuss-Winkler Model in routine practice is that its governing equation (4.20) can be readily combined with the governing equation of a foundation element (beam or plate) with subgrade support so that the subgrade reaction, p, is eliminated as an explicit problem variable. When the matrix method is used for structural analysis as is generally the case nowadays, it is easy to accommodate spatially varying values of the coefficient of subgrade reaction, k_1, so that the Pseudo-Coupled Concept strategies discussed in the preceding sections can be readily implemented if desired.

It turns out that there is another strategy that can be used to overcome the lack of inherent lack of spring coupling in the Fuss-Winkler Model. This alternative strategy was perceived and developed by the writer circa 1990. For reasons that will become clear, this alternative analytical strategy is called the *Beam-Column Analogy*. The attractions of this alternative methodology are that it actually incorporates the theoretical effects of spring coupling (and thus does not require any of the empirical assumptions or approximations of the Pseudo-Coupled Concept) while maintaining the analytical simplicity of the Fuss-Winkler Model.

The underlying concepts and theoretical development of the Beam-Column Analogy were presented in Horvath (1993e, with important corrections in 1993f). The key portions of this work are repeated here along with some previously unpublished insights and extensions that the writer developed subsequent to the original publication in 1993.

For simplicity, the presentation herein is limited to the 2-D problem of a beam-type foundation element with subgrade support. The concepts can readily be extended to the more general 3-D problem of a plate-type foundation element with subgrade support.

To begin with, Equation 4.21 that defines the behavior of a beam of constant flexural stiffness, EI, with subgrade support, p, is restated here for ease of reference and with simplified notation for the total derivative:

$$EIw^{IV} + p = q. \tag{4.28}$$

As with the Pseudo-Coupled Concept, the desire is to eliminate the subgrade reaction, p, as an explicit problem variable. However, unlike with the Pseudo-Coupled Concept that replaced p by using the governing equation of the Fuss-Winkler Model (Equation 4.20), in this case the governing equation of a multiple-parameter model that inherently incorporate spring coupling in its mathematical behavior is used. This, then, is the novel element of the Beam-Column Analogy as noted previously, i.e. the incorporation of spring coupling mathematically and without any arbitrary assumption or guesswork.

It should be apparent that the choices for a multiple-parameter model to use here are very limited. This is because the candidate model cannot have a governing differential equation with derivatives of p as this would negate the simplicity necessary for the overall process. This means that only models of Order H^2 in Figure 4.1 can be used.

In the writer's original circa-1990 research and concomitant publication (Horvath 1993e), the multiple-parameter mechanical model used was the Pasternak/Loof Hypothesis as envisaged by A. D. Kerr with a shear-only layer providing the shear coupling (Equation 4.15). Here, the equation is simplified for 2-D behavior:

$$p = k_1 w - g_1 w^{II}. \tag{4.29}$$

However, for reasons that will become clear, it is much more advantageous to use the Filonenko-Borodich Model (Equation 4.13) that is mathematically identical as can be seen from Tables 4.3 and 4.4. In 2-D, the governing differential equation of this model simplifies to:

$$p = k_1 w - T_1 w^{II}.$$

(4.30)

Substituting Equation 4.30 into Equation 4.28 yields the governing equation of a beam supported on a multiple-parameter subgrade of Order H^2 which, as can be seen in Table 4.4, is the lowest order subgrade with inherent spring coupling in its formulation:

$$EIw^{IV} + (k_1 w - T_1 w^{II}) = q.$$

(4.31)

Rearranging the terms in Equation 4.31 yields:

$$EIw^{IV} - T_1 w^{II} + k_1 w = q$$

(4.32)

which is recognized as being identical in form to the equation of a beam-column (Equation 2.8) that in this case is subjected to a tensile axial force of magnitude T_1 and with the simpler Fuss-Winkler subgrade support that is readily modeled in commercially available structural analysis software. In essence, by transforming the beam into a beam-column and transferring the subgrade spring-coupling effects into structural effects the subgrade is transformed from a multiple-parameter model into the simpler single-parameter Fuss-Winkler Model.

Using the alternative matrix-method formulation, Equation 4.32 becomes:

$$[S]\{d\} + [k_1]\{d\} = \{q\}$$

(4.33)

which simplifies to:

$$[S']\{d\} = \{q\}$$

(4.34)

where $[S']$ is the modified or enhanced stiffness matrix and is defined as follows:

$$[S'] = [S + k_1].$$

(4.35)

Note that in this case, [S] must be the stiffness matrix of a beam-column (Equation 2.9).

As noted in Chapter 2, the stiffness matrix in general provides interesting qualitative insight into fundamental aspects and implications of structural behavior in many cases as it makes visually obvious what the effect of various problem parameters such as shear effects and axial force have on the apparent (as opposed to inherent) flexural stiffness of a beam.

With reference to Equation 2.9 that is the stiffness matrix of a beam-column, in the case of the Beam-Column Analogy it is clear that the pseudo-axial force, T_1, acting on the beam tends to _increase_ the apparent flexural stiffness of the beam because this force is tensile in sense and thus negative in sign as the geotechnical sign convention was used for the structural derivations shown in Chapter 2. The clear implication is that the consideration of spring-coupling in mechanical subgrade models makes the foundation element supported by the subgrade appear to be flexurally stiffer compared to the same foundation element supported on a single-parameter (Fuss-Winkler) subgrade.

In summary up to this point, a beam supported on a multiple-parameter mechanical subgrade that _inherently_ incorporates spring coupling in its formulation (albeit in the simplest way possible) is mathematically equivalent to a beam-column supported on a single-parameter (Fuss-Winkler) subgrade with a spatially uniform coefficient of subgrade reaction, k_1. Thus, all the analytical simplicity of the Fuss-Winkler Model is retained but with the benefit of at least a first-order approximation of spring-coupling effects. As discussed in Chapter 2, the only caveat is that a nonlinear structural analysis is required in order to realize the inherent nonlinear effects that are reflected in the derivation of the beam-column equation. However, nowadays this is a trivial requirement to satisfy, even in routine practice.

There are several additional issues of relevance to note concerning the Beam-Column Analogy. First, as noted in Horvath (1993e), as an alternative to using a mechanical subgrade model of Order H^2 for the subgrade reaction, p, in Equation 4.28, a subgrade model of the same Order but based on a simplified elastic continuum could be used. As can be seen in Table 4.2, there are three candidate equations that could be used for this purpose.

In Horvath (1993e), only the PTSC (Equation 4.8) based on the writer's extension of Reissner's simplification concepts was noted. The equation for the PTSC simplified for 2-D behavior is:

$$p = \left(\frac{E}{H}\right)w - \left(\frac{GH}{2}\right)w^{II} .$$ (4.36)

When coupled with the equation for a beam with subgrade support (Equation 4.28), the final result is:

$$EIw^{IV} - \left(\frac{GH}{2}\right)w^{II} + \left(\frac{E}{H}\right)w = q$$ (4.37)

which compares to Equation 4.32 for the formulation using a mechanical subgrade model. Note that in this case $(GH/2)$ becomes the fictitious tensile force applied to the beam-column and (E/H) is the equivalent subgrade 'spring' stiffness.

For the sake of completeness, it is noted that a Rhines-Solution Simplified Continuum of Order H^2 could be used in lieu of the PTSC. The relevant equation for this model simplified for 2-D conditions is:

$$p = \left[\frac{E}{H(1 - v^2)}\right]w - \left[\frac{EH}{3(1 - v^2)}\right]w^{II} .$$ (4.38)

When coupled with the equation for a beam with subgrade support (Equation 4.28), the final result is:

$$EIw^{IV} - \left[\frac{EH}{3(1 - v^2)}\right]w^{II} + \left[\frac{E}{H(1 - v^2)}\right]w = q .$$ (4.39)

Note that although the constant coefficients in this equation are broadly similar to those in Equation 4.37 for a Reissner-type simplified continuum, there are differences that would produce different numerical results for the same problem application. This is not surprising as Rhines solved a structurally more restrictive problem compared to Reissner as discussed earlier in this chapter.

To complete the discussion of the Beam-Column Analogy, as discussed in Horvath (1993e), there are two pragmatic issues that must be addressed in order to use the Beam-Column Analogy in a practical application. The first is the combination of parameter implementation in commercially available structural analysis software and parameter evaluation.

The second issue is boundary conditions related to the subgrade component[42]. As discussed earlier in this chapter with specific regard to subgrade models of Order H^3 such as the RSC that are one Order higher than the Beam-Column Analogy, this is an issue that arises with all multiple-parameter models, whether developed based on simplifications to an elastic continuum or an assemblage of mechanical elements. Specifically, behavior beyond the edge of the subgrade reaction, p, must be considered explicitly as there will still be settlements, w, even though the subgrade reaction, p, is zero. As noted previously, this issue does not arise with any single-parameter model (i.e. the F/WTSC or Fuss-Winkler mechanical model) because there is no subgrade settlement, w, when $p = 0$.

Considering parameter implementation and assessment first, this issue was addressed in Horvath (1993e) and although the term was not used at that time the discussion actually set the stage for what are now formally called *hybrid subgrade models* that combine the best elements of the two traditional subgrade model types (i.e. simplified continuum and mechanical). Hybrid subgrade models are a central feature of this monograph as they represent the current state of art in subgrade modeling. Consequently, they are discussed formally later in this chapter but the basic elements as they apply to the Beam-Column Analogy are discussed here.

To begin with, a significant benefit of mechanical subgrade models is that the element types used to formulate such models are, in general, straightforward and easy to model in commercially available structural analysis software compared to the parameters (E, G, ν, H) that appear with continuum-based subgrade models. The one caveat is that a deformable tensioned membrane[43] is much easier to model structurally than the mathematically equivalent shear-only layer which is why in the above derivation of the Beam-Column Analogy the Filonenko-Borodich Model was preferred over A. D. Kerr's visualization of the Pasternak/Loof Hypothesis that the writer had used originally in Horvath (1993e).

The downside of mechanical models is, of course, the eternal question of how to evaluate the constant coefficients that are the model parameters (k_1 and T_1 in this case with reference to Equation 4.32). This is where the benefit of hybridization between mechanical and continuum-based models comes in. It is obvious that the final governing equations of the Beam-Column Analogy (Equation 4.32 for the mechanical model version and Equation 4.37 for the Reissner-type simplified-continuum model version[44]) are mathematically identical. This means that $k_1 = E/H$ and $T_1 = GH/2$. There was some discussion in Horvath (1993e) on how to evaluate E, G, and H using the case history example presented in that paper. A more-extensive and updated discussion of this important topic is presented in Chapter 5.

This leaves the question of subgrade boundary conditions. This topic was discussed in some detail in Horvath (1993e) but a summary is presented here.

With a subgrade model of Order H^2, there is only one subgrade-related boundary condition and it directly involves w', the first derivative of subgrade displacement, w, with respect to the horizontal (x) axis. This boundary condition thus involves assumptions related to the vertical shear force within the subgrade that results from the spring coupling inherent in subgrade models of this Order as this shear force is linearly proportional to w'. In any event, one of the following two choices must be made:

---

[42] As noted in Chapter 3, there are also the usual boundary conditions related to the foundation element. These are well known and are not discussed further in this monograph.
[43] Note that a nonlinear analysis is always required in order to properly model the deformability requirement.
[44] Alternatively, Equation 4.39 for a Rhines-Solution Simplified Continuum could be used.

- Both w and w' are continuous at the ends of the beam.

- Only w is continuous at the ends of the beam, with w' discontinuous which produces a kink in the settlement pattern at the ends of the beam.

In view of the discussion earlier in this chapter with regard to similar choices for the Order H^3 subgrade models, the latter would intuitively seem to be the better choice. However, as discussed in Horvath (1993e) and illustrated there using a case history application, the former appears to be the better choice for the Beam-Column Analogy.

To reconcile this unexpected (to the writer at least) outcome, it is relevant to bring up a subtle point that the writer did not raise in Horvath (1993e) that impacts the boundary condition issue for the Beam-Column Analogy. This is the fact that the Beam-Column Analogy is not exactly the same problem as an ordinary beam supported on a subgrade of the same Order (H^2). This is illustrated in Figure 4.3 which shows the Beam-Column Analogy vs. a beam supported on a Filonenko-Borodich Model subgrade.

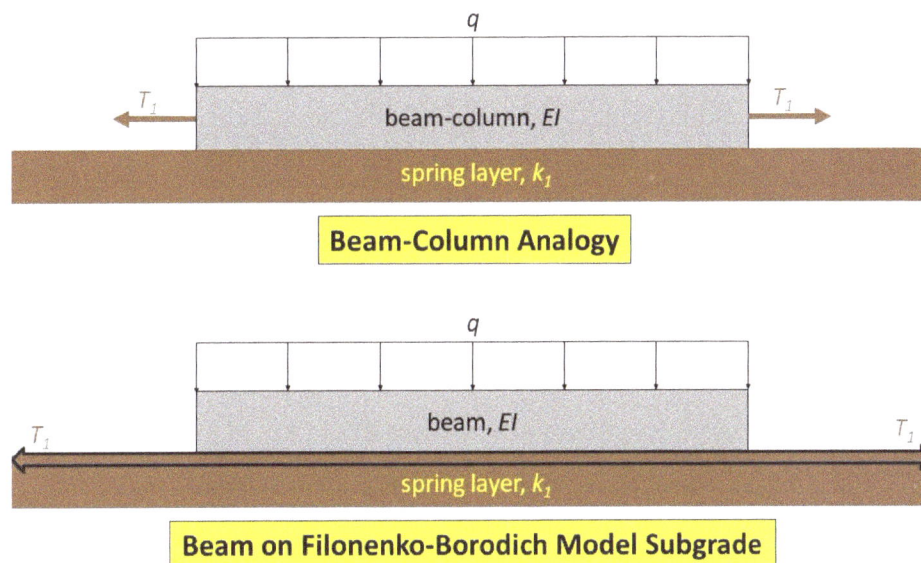

Figure 4.3. Beam-Column Analogy vs. Beam on Filonenko-Borodich Model Subgrade.

As can be seen in this figure, with the Beam-Column Analogy a multiple-parameter subgrade model (specifically, the Filonenko-Borodich mechanical model consisting of a deformed/deformable tensioned membrane over a spring layer) is incorporated into the problem but only within the limits of the foundation element (beam in this case). This is because the membrane component is included only within the limits of the beam by virtue of the assumed beam-column behavior. However, beyond the limits of the beam-column the subgrade is that of the spring-only Fuss-Winkler Model.

On the other hand, with a beam without an axial force modeled over a complete and separate Filonenko-Borodich Model subgrade, the tensioned-membrane effects extend beyond the limits of the beam. Note that in this case, the structural analyst would have to explicitly and separately model a beam without axial forces and a deformable tensioned membrane over a spring layer as opposed to simply a beam-column over a spring layer as with the Beam-Column Analogy.

Thus, the two problems shown in Figure 4.3 are subtly different as to what happens beyond the limits of the foundation element (beam in this case). Thus, in retrospect it is not surprising that the subgrade boundary conditions at the edges of the foundation element (ends of the beam in this case) would be different as well. This emphasizes the fact that the subgrade boundary condition alternatives for subgrade models should always be independently evaluated for a given model as to which conditions provide the best correlation with typical foundation behavior.

In conclusion, at the time the writer envisaged and developed the Beam-Column Analogy circa 1990 it was a clever way to incorporate at least first-order spring-coupling effects into SSI applications given the more-limited (relative to the present) capabilities of commercially available structural analysis software of that timeframe. However, nowadays it would not be a burden to model the same problem more completely by incorporating the spring-coupling effects as a separate element as opposed to incorporating these effects incompletely within the foundation element, both as illustrated in Figure 4.3. Thus, while at one time the Beam-Column Analogy was a practical and pragmatic means to an end to incorporate a subgrade model with inherent spring coupling, time and technology have rendered it more of an intellectual curiosity and technological dead end in the writer's opinion.

4.7.6 Synthesis

Tables 4.5 and 4.6 contain a synthesis of all mechanical models, both single- and multiple-parameter, that have been identified by the writer to date. As noted earlier in this chapter, a body of liquid could be substituted for a spring layer in any of these models but is only noted explicitly for the Hertz Model as this was apparently an explicit assumption of that model.

Table 4.5. Mechanical Elements for All Mechanical Subgrade Models.

Order	Subgrade model (in groupings of mathematically identical models)	Mechanical elements used to visualize model (as viewed from top to bottom)
H^1	Fuss-Winkler	springs
H^2	Filonenko-Borodich	deformed tensioned membrane + springs
	Pasternak's Hypothesis Loof's Hypothesis	shear layer + springs
H^3	Kerr (aka Modified Pasternak)	springs + shear layer + springs
	Horvath-Colasanti (aka Modified Kerr)	springs + deformed tensioned membrane + springs
-	Hertz	plate + liquid
	Haber-Schaim (aka Modified Hertz)	plate + springs
-	Hetényi	springs + plate + springs
H^5	Rhines	springs + plate + shear layer + springs

Table 4.6. Mathematical Hierarchy of All Mechanical Subgrade Models.

Order	Subgrade Model (in groupings of mathematically identical models)	derivative order of $p(x,y)$			derivative order of $w(x,y)$		
		0	2	4	0	2	4
H^1	Fuss-Winkler						
H^2	Filonenko-Borodich Pasternak's Hypothesis Loof's Hypothesis						
H^3	Kerr (aka Modified Pasternak) Horvath-Colasanti (aka Modified Kerr)						
-	Hertz Haber-Schaim (aka Modified Hertz)						
-	Hetényi						
H^5	Rhines						

4.8 HYBRID SUBGRADE MODELS

4.8.1 Generic Concept and Implementation

The overall goal of what the writer has defined as hybrid subgrade models is to create a hybridized or composite model that, on the one hand, is easily implemented into commercially available structural analysis software and, on the other hand, has constant coefficients defining model behavior that can be logically quantified on a site- and application-specific basis using conventional geotechnical site characterization strategies. This goal is achieved by selectively using components of both mechanical and simplified-continuum models and then combing them into one, synergistic analytical methodology. Specifically, selected mechanical elements are used to visualize the subgrade model for structural-modeling purposes and the parameters (k, T, etc.) that define the behavior of these mechanical elements are expressed in terms of the elastic parameters and geometry of the subgrade. The required subgrade boundary conditions are also defined in the same manner.

With regard to the mechanical-element types used in this process, in principle only axial springs and deformable tensioned membranes are required for the hybrid subgrade models identified to date that are noted in the following tables. Shear-only layers, while easy to conceptualize and visualize, should be avoided as they are more problematic to model in software compared to the mathematically equivalent deformable tensioned membrane.

The only mandatory requirement of the hybridization process is that the mechanical and simplified-continuum models that are combined must have the same form in their respective governing equations in terms of the derivative orders of the subgrade reaction, p, and subgrade settlement, w. To illustrate the potential model combinations that are available, in principle at least, for hybridization, Table 4.7 combines results shown previously in Tables 4.2 and 4.6 for simplified-continuum and mechanical subgrade models, respectively, that have been identified to date by the writer. Note that Table 4.7 is artificially limited to models of Order H^5. In reality, this table could be extended indefinitely as there is no inherent limit to either mechanical models (as is obvious in Figure 4.1) or simplified-continuum models (Equation 4.1).

96

Table 4.7. Mathematical Hierarchy and Synthesis of Simplified-Continuum and Mechanical Subgrade Models.

Order	Subgrade Model (in groupings of mathematically identical models)	derivative order of $p(x,y)$			derivative order of $w(x,y)$		
		0	2	4	0	2	4
H^1	**mechanical** Fuss-Winkler **simplified continuum** Reissner-type (Fuss/Winkler-Type Simplified Continuum) Rhines-Solution Simplified Continuum of Order H^1						
H^2	**mechanical** Filonenko-Borodich Pasternak's Hypothesis Loof's Hypothesis **simplified continuum** Reissner-type (Pasternak-Type Simplified Continuum) Vlasov-Leont'ev/1 layer Rhines-Solution Simplified Continuum of Order H^2						
H^3	**mechanical** Kerr (aka Modified Pasternak) Horvath-Colasanti (aka Modified Kerr) **simplified continuum** Reissner-type (Reissner Simplified Continuum) Vlasov-Leont'ev/2 layers Rhines-Solution Simplified Continuum of Order H^3						
-	**mechanical** Hertz Haber-Schaim (aka Modified Hertz) **simplified continuum** Rhines-Solution Simplified Continuum						
-	**mechanical** Hetényi **simplified continuum** Rhines-Solution Simplified Continuum						
H^4	**mechanical** <unnamed - see Figure 4.1> **simplified continuum** Rhines-Solution Simplified Continuum of Order H^4						
H^5	**mechanical** Rhines **simplified continuum** Rhines-Solution Simplified Continuum of Order H^5						

Subgrade Modeling and Models in Foundation Engineering
John S. Horvath, Ph.D., P.E., Life Member.ASCE

Table 4.8 extends Table 4.7 to explicitly show the three hybrid subgrade models that have been formally identified and developed by the writer to date[45]. The correlation for the overall simplest hybrid model (of Order H^1) based on the Fuss-Winkler Hypothesis with a constant coefficient (Equation 4.9 for the Fuss-Winkler Type Simplified Continuum (F/WTSC) based on the writer's extension of Reissner's simplification concepts and Equation 4.20 for the Fuss-Winkler mechanical model based on the spring analogy) is trivial, i.e. $k_1 = E/H$ and with no subgrade boundary conditions to consider.

The correlation for the simplest hybrid model (of Order H^2) that explicitly incorporates spring coupling in the mathematical formulation combines the Filonenko-Borodich mechanical model and the Pasternak-Type Simplified Continuum (PTSC) based on the writer's extension of Reissner's simplification concepts. The general outcomes of this hybridization are similar to that of the Beam-Column Analogy that was discussed earlier in this chapter.

The most advanced hybrid model (of Order H^3) developed to date combines the Horvath-Colasanti mechanical model and Reissner's original simplified-continuum model (RSC). This hybrid model, which is called the *Horvath-Colasanti/Reissner (H-C/R) Model*[46], was first presented in Horvath and Colasanti (2011a), with the subgrade boundary condition development presented in Colasanti and Horvath (2011). For ease of reference, relevant details of the H-C/R Model are contained in Appendix C of this monograph.

The H-C/R Model represents the current state of art for subgrade modeling. As such, Chapters 5 and 6 are devoted to addressing how this model can be implemented into routine practice using existing technology that is widely available to all design professionals.

However, that is not to say that geotechnology should stand still. As can be seen in Table 4.8, hybrid models of higher order await development. This includes the dual needs of developing correlations between the mechanical and simplified-continuum model parameters as well as developing subgrade boundary conditions that can be implemented into commercially available structural analysis software.

4.8.2 Limitation

The only limitation on implementing the hybrid subgrade model concept is that it can only be used for SSI applications where a subgrade model is used to model the behavior of only the subgrade. Historically, this was always the case and still is the predominant application of subgrade models to the present. However, beginning in the 1990s there was a burst of research and concomitant publication devoted to alternative composite uses of subgrade models. This was for SSI applications involving traditional planar (2-D) geosynthetics such as geogrids, geomembranes, and geotextiles in applications involving the geosynthetic function of tensile reinforcement. In such applications, which are discussed in the following section, multiple-parameter mechanical subgrade models are used but the parameters in the model represent the combined behavior of all components (i.e. geosynthetic and soil subgrade) of the reinforced soil system. Consequently, it is not appropriate to try to match the parameters of a mechanical subgrade model in such an application with the parameters of a simplified-continuum subgrade model of the same Order as the continuum-model parameters only apply to the soil-subgrade component.

---

[45] Note that the previously discussed Beam-Column Analogy does not appear in this table. This is because it is not a complete, 'true' hybrid subgrade model as illustrated in Figure 4.3 although it does make use of hybridization concepts in order to evaluate equation coefficients in terms of the elastic parameters and thickness of the subgrade.
[46] Earlier publications refer to this as the *Modified Kerr-Reissner* (MK-R) *Model*.

Table 4.8. Mathematical Hierarchy of Hybrid Subgrade Models Identified to Date.

Order	Subgrade Model (in groupings of mathematically identical models)	derivative order of p(x,y)			derivative order of w(x,y)		
		0	2	4	0	2	4
H^1	**mechanical** Fuss-Winkler **simplified continuum** Reissner-type (Fuss/Winkler-Type Simplified Continuum) Rhines-Solution Simplified Continuum of Order H^1 **hybrid** Fuss-Winkler						
H^2	**mechanical** Filonenko-Borodich Pasternak's Hypothesis Loof's Hypothesis **simplified continuum** Reissner-type (Pasternak-Type Simplified Continuum) Vlasov-Leont'ev/1 layer Rhines-Solution Simplified Continuum of Order H^2 **hybrid** Filonenko-Borodich/PTSC						
H^3	**mechanical** Kerr (aka Modified Pasternak) Horvath-Colasanti (aka Modified Kerr) **simplified continuum** Reissner-type (Reissner Simplified Continuum) Vlasov-Leont'ev/2 layers Rhines-Solution Simplified Continuum of Order H^3 **hybrid** Horvath-Colasanti/Reissner (aka Modified Kerr-Reissner)						
-	**mechanical** Hertz Haber-Schaim (aka Modified Hertz) **simplified continuum** Rhines-Solution Simplified Continuum						
-	**mechanical** Hetényi **simplified continuum** Rhines-Solution Simplified Continuum						
H^4	**mechanical** <unnamed - see Figure 4.1> **simplified continuum** Rhines-Solution Simplified Continuum of Order H^4						
H^5	**mechanical** Rhines **simplified continuum** Rhines-Solution Simplified Continuum of Order H^5						

4.9 ALTERNATIVE COMPOSITE-BEHAVIOR USES OF SUBGRADE MODELS

4.9.1 Basic Concept

During the 1990s, there was a burst of research and concomitant publication on the subject of analyzing geosynthetically reinforced soil as a displacement-based SSI problem using subgrade models as opposed to using traditional strength-based limit-equilibrium analytical methodologies. Some relevant publications from that timeframe (this is by no means an exhaustive or complete list) include Douglas (1995), Ghosh and Madhav (1994a, 1994b, 1994c), Horvath (1994), Shukla and Chandra (1994a, 1994b, 1994c, 1995), and Yin (1997a, 1997b). As can be seen, many of these papers were by the same authors and their individual papers each tended to address one narrow aspect of the same overall problem, that of embedding geosynthetic tensile reinforcement within a coarse-grain soil fill layer that was placed over a pre-existing fine-grain soil subgrade.

There are two notable aspects to this collective body of work. One is the development and use in some cases of time-dependent subgrade models, with the temporal effects only used to model primary consolidation of the fine-grain soil subgrade. The fact that polymeric geosynthetic tensile reinforcement also has significant time-dependent behavior was not modeled explicitly.

However, the more significant of the two issues concerning these publications is the use of the subgrade models (mechanical models in all cases) to model not only the soil components but the embedded geosynthetic tensile reinforcement as well. While there is nothing wrong with this inherently, it does preclude application of the hybridization process for the reasons discussed previously. As a result, one is forced to work with the mechanical subgrade models as-is with all the well-known, attendant difficulties of trying to quantify the abstract parameters (k, g, T) that define the behavior of such subgrade models. This fact alone severely limits the practical utility of using a displacement-based SSI approach to modeling geosynthetically reinforced soil in routine practice, at least when the subgrade model is used to model the composite geosynthetic + subgrade behavior.

4.9.2 Alternative Approach

For the sake of completeness, it should be noted that there is an alternative approach to modeling the generic problem of geosynthetically reinforced soil as a displacement-based SSI problem. Of significance is that this alternative approach does allow developing hybrid subgrade models and thus has much greater potential for development and use in both practice and research. To do so, the use of the subgrade model simply needs to be limited to the soil subgrade, with the geosynthetic tensile reinforcement modeled separately as an equivalent foundation element.

This approach was illustrated using a simple proof-of-concept example in Horvath and Colasanti (2011b). Of note is that the results obtained using the aforementioned Horvath-Colasanti/Reissner hybrid subgrade model (which was called the MK-R Model in the paper) compared very favorably for all cases studied to results obtained using an 'exact' FE analysis of a continuum. Although the writer has not pursued this application of subgrade models and SSI analysis beyond the proof-of-concept stage, it remains a promising analytical approach awaiting further development.

This page intentionally left blank.

Chapter 5

Hybrid Subgrade Model Parameter Development in Practice

5.1 INTRODUCTION AND OVERVIEW

The preceding chapter showed that when a subgrade model is used to represent subgrade behavior only (which covers the vast majority of SSI applications) it is possible to use a hybrid subgrade model that combines the best attributes of separate mechanical and simplified-continuum subgrade models into one practical analytical tool.

It is the writer's opinion that a hybrid model should always be the subgrade model of choice nowadays, at least whenever generic structural analysis software is used for a given SSI application. Furthermore, a multiple-parameter hybrid model such as the Horvath-Colasanti/Reissner (H-C/R) Model should always be used rather than using the hybrid version of the single-parameter Fuss-Winkler Model with some version of the Pseudo-Coupled Concept (it should be clear by this point that using the Fuss-Winkler Model with a constant coefficient of subgrade reaction, k_1, is simply no longer defensible in any application).

As an aside, there are and will likely continue to be exceptions to these general guidelines of using a multiple-parameter hybrid subgrade model. For example, it is likely that the *p-y* Curve method for laterally loaded deep foundations, which is essentially the single-parameter Fuss-Winkler Model with the Pseudo-Coupled Concept embedded in application-specific software, will continue to be used in both practice and research even though a more theoretically rigorous multiple-parameter simplified-continuum subgrade model for this application was identified more than 30 years ago (Horvath 1984a). The reason is that the *p-y* Curve method is a well-established analytical tool that has achieved wide acceptability after over six decades of refinement and use worldwide in proprietary software that continues to evolve over time.

With the suggested use of multiple-parameter hybrid models for generic applications with commercially available structural analysis software in mind, the remainder of this monograph is devoted to illustrating how this can be done using everyday technology available to design professionals. Specific attention is given to use of the H-C/R Model of Order H^3 as it is the most advanced hybrid model that not only has been explicitly identified to date but for which both the necessary parameter correlations between its mechanical and simplified-continuum components as well as subgrade boundary conditions have been researched and developed as discussed in Appendix C. However, much of the following discussion is sufficiently broad and generic so that the material presented could be used with other hybrid models, including those of even higher Order and thus greater accuracy that are yet to be identified and developed.

The remainder of this chapter is devoted to illustrating the process for developing the necessary parameters for the simplified-continuum component of a hybrid model so that the parameters defining the behavior of the companion mechanical component, including the subgrade boundary conditions, can be quantified. The common application of a mat foundation, which has been used as the exemplar SSI application throughout this monograph, is used for this purpose. The following Chapter 6 illustrates several mat foundation case histories to which the process illustrated in this chapter was applied and compares forecast vs. measured results.

5.2 PROCESS OVERVIEW

One of the several benefits of using hybrid subgrade models is that it transforms quantifying the abstract mechanical-model parameters (springs, etc.) used in structural analysis software into a rational process, something that has been lacking since subgrade models began to evolve two centuries ago. Figure 5.1 shows the four steps involved in the process in an idealized fashion. Note that all depths, z, and dimensions are relative to foundation level, i.e. the assumed planar interface between the foundation element and subgrade.

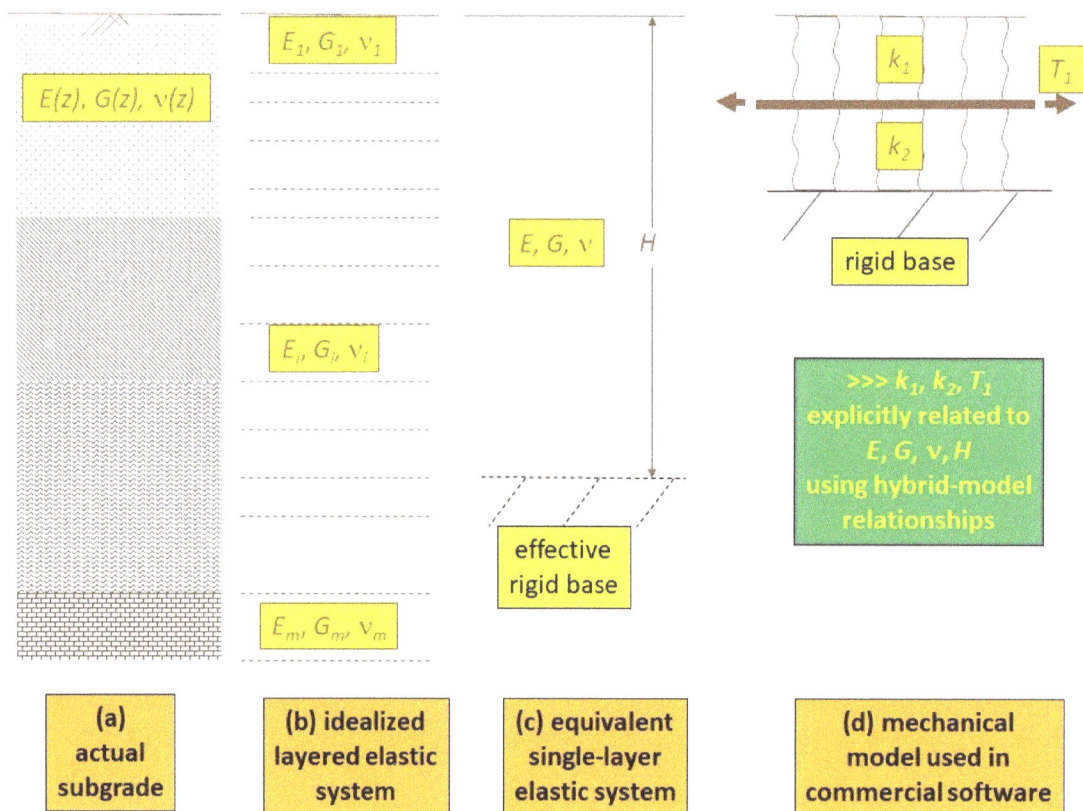

Figure 5.1. Process Overview for Quantifying Hybrid Subgrade Model Parameters.

The first three steps in the overall process of using a hybrid subgrade model in practice involve evaluating the parameters of the simplified-continuum component that is used to create the hybrid model (the Reissner Simplified Continuum (RSC) in this case):

1. Perform site characterization to define the stratigraphy and relevant engineering properties of the actual subgrade (Figure 5.1a).

2. Divide the actual subgrade into an idealized, artificially layered system (Figure 5.1b).

3. Convert the layered system into an equivalent single-layer system with, most importantly, a depth, H, to an effective, artificial rigid base. This depth is often referred to

synonymously in the published literature as the *relative depth of influence*, *significant stressed depth*, or similar term.

The following sections elaborate on each of these steps.

The fourth and final process step (Figure 5.1d) is to evaluate the mechanical parameters for the relevant mechanical component (the Horvath-Colasanti Model in this case) that is implemented into the commercially available structural analysis software used to perform an analysis. This parameter evaluation uses pre-established correlations with the relevant simplified-continuum component (the RSC Model in this case which are given in Appendix C). The necessary subgrade boundary conditions are evaluated as well (the resulting additional edge springs for the Horvath-Colasanti Model are also given in Appendix C).

Examples of actual implementation and outcomes of this overall four-step process into a specific brand (*ANSYS*) and version (11.0) of commercially available structural analysis software for several case histories are presented in Chapter 6.

5.3 SITE CHARACTERIZATION

SSI applications where subgrade models are used as the primary analytical tool are generally governed by subgrade displacements as opposed to subgrade strength. Consequently, site characterization studies performed in support of subgrade modeling for SSI applications should focus on defining site- <u>and</u> application-specific conditions that contribute to displacements (settlement in the case of mat foundations as have been used as the exemplar of SSI applications throughout this monograph). Note that the application-specific aspect has been intentionally emphasized here as the site characterization study undertaken for, say, a warehouse slab-on-grade vs. a mat supported high-rise building at the same site should be completely different.

With this overarching guideline in mind and assuming that settlement is the primary displacement parameter of interest, the key requirements for a site- and project-specific site characterization assessment for SSI applications are defining the:

- actual subsurface stratigraphy and

- geomaterial stiffnesses,

both to a depth that is at or below that expected to contribute significantly to settlement of the proposed structure ('rigid base' in the context of Figure 5.1c and the following discussions).

With regard to the latter requirement, although both the Young's and shear moduli are shown in Figure 5.1a, in reality on a given project it is only one or the other that is typically estimated directly as part of the site characterization process. Historically, the emphasis was on Young's modulus, E, but as site investigations in recent years have increasingly included in-situ measurement of shear wave velocities, the trend has been toward using the shear modulus, G.

Note that this stiffness, whether E or G, must be relevant to both the stress and strain levels that are expected to be imposed by the proposed structure. As is well known, soil moduli are, in general, never constant in magnitude but are both stress and stress-level dependent. This aspect of soil behavior can be very challenging to deal with properly and requires careful consideration on the part of the design professional. This is because the wide range of laboratory and in-situ testing technologies available today in routine practice produce estimates of moduli that correspond to a wide range of stress, stress level, and strain conditions. For example, the small-strain shear modulus, G_{max} or G_o[47], produced by various in-situ testing devices typically must be degraded (reduced) to a value of shear modulus that corresponds to the operative strain level that is expected in a given application. Furthermore, to the extent that stress state, especially *yield stress*[48], affects the relevant stiffness of the subgrade this must be determined as well.

In-situ testing is now recognized as being an essential tool in accomplishing the goals of site characterization for SSI applications. This includes a wide spectrum of devices:

- <u>invasive</u> tools that include penetrometers with essentially continuous data recording such as the *piezocone* (CPTu) as well as devices such as the *dilatometer* (DMT) and *pressuremeter* (PMT) that are typically used to produce data at discrete depth intervals;

- <u>non-invasive</u> techniques such as *seismic refraction* and *spectral analysis of surface waves* (SASW) that involve generating sub-acoustic[49] surface waves and making surface measurements of responses to the generated waves; and

- hybrid technologies such as the *seismic piezocone* (sCPTu) that use an invasive tool to record sub-acoustic waves generated at the surface.

While each of these tools has its place in routine practice depending on specific ground conditions, the writer has found that the current variants (CPTu and sCPTu) of the basic *cone penetrometer* (CPT) are the most cost-effective and useful tools for SSI applications and subgrade models when ground conditions permit their use. This is because the basic CPT is one of the oldest in-situ testing devices used by geotechnical engineers, second only the Standard Penetration Test (SPT). The CPT dates back to the 1930s and thus we are approaching a century of use and refinement of the device itself.

More importantly, there have been many decades of developing empirical correlations between the parameters (q_c-f_s-u_2) measured using the CPTu/sCPTu and soil properties. In recent years, most the of research into these correlations has been performed by Prof. Paul Mayne (e.g. Mayne 2006, 2007) and Dr. Peter K. Robertson (e.g. Robertson and Cabal 2015). As a result, the writer has found that it is possible to extract dozens of soil properties and other relevant parameters from CPTu soundings, especially when some proportion of soundings on a project are performed using the sCPTu. This is part of the aforementioned recent trend toward using in-situ testing methodologies that produce and measure shear-wave velocities and then use these velocities to estimate the small-strain shear modulus, G_{max} or G_o.

---

[47] These parameter notations are used synonymously in the published literature although there are apparently regional preferences. For example, G_{max} tends to be more common in the U.S. whereas G_o is more common in Canada and Europe. Also, in some publications (e.g. Horvath 2016a, 2016b) there is an intentional use of one (G_{max} in this case) over the other in order to avoid a conflict with another, completely different parameter defined by others that uses the same notation (G_o in this case).

[48] As noted in Chapter 3, the parameter of yield stress is also referred to by a wide variety of alternative terms such as *maximum past effective stress/pressure, preconsolidation stress/pressure,* and *natural prestress* with concomitant variations in notation. The term yield stress is preferred as it is now recognized that the stiffness behavior of geomaterials is not always related to actual past loading history but may be influenced by weathering and related microstructure as in the case of laterites and other residual soils and heavily weathered bedrock.

[49] Note that surface generation of the wave itself is typically a human-audible process.

5.4 ARTIFICIAL LAYER DETERMINATION

The second step in the overall process of using a hybrid subgrade model in practice involves creating a system of some arbitrary number, m, of subgrade layers of arbitrary thicknesses (Figure 5.1b). The focus of this step should be on creating layers that each has constant stiffness, either E or G. The degree of professional judgment and subjectivity that goes into this process can vary widely and depends primarily on:

- the analytical tool used to generate the stiffness profile in the first step (Figure 5.1a) and

- whether the averaging process that comprises the next (third) step in the process will be done manually by scaling off some published chart solution or automated into a computer-based solution.

With regard to the first item, if, for example, CPTu/sCPTu data are used directly for empirical correlation with E or G then the depth-wise data spacing may be as close as 20 millimetres (0.8 in). In that case, the determination of artificial layering may resort to being a highly subjective exercise as was the case with Schmertmann's visual/manual procedure that was used in his original CPT-based methodology for settlement analysis of footings (Schmertmann 1978).

On the other hand, if the moduli data are obtained at relatively widely spaced depth intervals using sCPTu shear-wave data, a DMT, or a PMT, then the data will generally be 1.5 metres (5 ft) apart or greater. In that case, it might be reasonable to simply assume that each piece of data defines an artificial layer, especially if an automated averaging methodology is used in the next step in the overall process.

5.5 EQUIVALENT SINGLE-LAYER DETERMINATION

5.5.1 Overview

The third step in the overall process of using a hybrid subgrade model in practice, and the last step discussed in this chapter, involves the determination of an equivalent homogeneous, isotropic, linear-elastic layer of finite thickness (Figure 5.1c). In many ways, this is the most crucial step in the overall process as it involves the generation of two distinct problem-specific constants that alone and together significantly influence the calculated values of the parameters for the simplified-continuum component of the hybrid model. In turn, these simplified-continuum parameters have a significant effect on the calculated values of the parameters for the mechanical component as well as subgrade boundary conditions of the hybrid model that are both critical inputs into the commercially available structural analysis software used to perform the SSI analysis.

The two problem-specific constants that are generated as the outcomes from this third step in the overall process are:

- H, the depth (relative to foundation level) to an equivalent 'rigid' base. This parameter is more accurately defined and visualized as the depth below which the subgrade is assumed to not contribute to the settlement of the foundation element being analyzed. As noted previously, there are other terms that are often used synonymously for this such as 'significant stressed depth'. However, the concept of an equivalent rigid base is consistent with concepts used in developing simplified-continuum subgrade models so is used here to maintain that conceptual linkage.

- The single-value modulus, E or G, that represents an average[50] of the moduli for each artificial layer shown in Figure 5.1b. Once this average modulus (e.g. G) is determined, the other modulus (e.g. E) is estimated using the theoretical relationship that links G and E. This is because E and G along with H are required to evaluate the parameters for the simplified-continuum component of the hybrid subgrade model used.

5.5.2 Estimation of Depth to Effective Rigid Base

5.5.2.1 Overview

As noted previously, within the context of SSI applications and hybrid subgrade models, the depth, H, to an effective rigid base as shown in Figure 5.1c is defined as the depth below foundation level at which settlements caused by subgrade loading from the foundation element (e.g. mat as has been the exemplar illustrative SSI application used throughout this monograph) can be taken to be zero. Thus, in any practical application this depth will be the lesser of the actual depth to a stratum of geomaterial that is substantially stiffer than overlying geomaterials or some equivalent depth determined by either a rule-of-thumb assumption or the outcome of some analytical methodology.

Experience has shown that a reasonably accurate determination of H is critical to the successful implementation of hybrid subgrade models in general and the H-C/R Model in particular. Consequently, this particular issue is discussed here in some detail. In Chapter 6 that presents several case histories using the H-C/R Model, examples of situations where H is assumed to be an actual 'rigid' layer as well as where it is an effective or imaginary one are illustrated and the impact on calculated results discussed.

It is important to note that this sensitivity of results to the assumed depth to rigid base should not be perceived as a unique disadvantage of either hybrid subgrade models in general or the H-C/R Model in particular. It is well known that that shallow-foundation settlement forecasts made using many analytical methodologies that have been proposed since the emergence of 'engineered' foundation engineering in the 1950s, especially those based on some aspect of the theory of linear elasticity as is the hybrid subgrade model concept, are sensitive to an assumed depth to rigid base (Schmertmann 1978, Mayne 2005). Thus, this is an issue that comes up in geotechnical applications using many other analytical methodologies. This is also illustrated in Chapter 6 for a case history application.

Because the determination of H is such an important issue for successful application of the hybrid subgrade model concept that is a central element of this monograph, over the years the writer has given particular research attention to developing a rational analytical methodology for its assessment.

5.5.2.2 Traditional Assumptions

Historically, rules of thumb have been used to define the depth to rigid base, especially in analytical methodologies geared toward forecasting the settlement of footings. By far the most common of these is the well-known '$H = 2B$' assumption where B is the width of the shallow-foundation element that is generally implied, if not stated explicitly, to be

[50] The specific kind of average is discussed subsequently in detail.

square in plan view. Even methods based largely on the theory of elasticity such as the well-known Schmertmann Method (Schmertmann 1978) have used such assumptions although Schmertmann did refine things somewhat by using a linear variation between $2B$ and $4B$ depending on the length-to-width (L/B) ratio of the foundation, i.e. =1 or = ∞ where $L/B \geq 10$ is generally taken to be $\approx \infty$.

5.5.2.3 Modern Perspectives

Research in recent decades has clearly demonstrated that such traditional rules of thumb concerning the relative depth below a shallow foundation (footing, mat, etc.) at which settlement caused by the foundation is for all practical purposes zero are simply not valid for the wide range (two-plus orders of magnitude) of widths that shallow foundations can exhibit in practice. To begin with, these rules of thumb were, in general, based on:

- theory-of-elasticity solutions for systems where Young's modulus is constant with depth (a situation not typical of most actual subgrades) and

- a consideration of only the vertical normal-stress <u>change</u> induced within the subgrade by the foundation element and the not <u>absolute</u> vertical stresses that include overburden stresses.

Some specific, seminal publications that have explored these issues are those of Burland and Burbidge[51] (1985), Berardi and Lancellotta (1991), and Charles (1996). These references have shown collectively and unequivocally that the relative depth, H/B, to an effective rigid base for settlement analysis purposes is never a simple constant (e.g. in the range of 2 to 4 depending on L/B as discussed in the preceding section), even for a given loaded-area shape in plan view (i.e. square, etc.). Rather, H/B is highly dependent on both the absolute width of the loaded area as well as the relative (to overburden stresses) magnitude of the applied load, with decreasing relative depth of influence, H/B, with increasing width of the loaded area and/or decreasing applied load. Berardi and Lancellotta indicated that there is also a secondary dependence of the H/B ratio on soil stiffness.

Discussing first the work of Burland and Burbidge, based on a composite of results from a theory-of-elasticity solution that assumed a modulus that increased with depth plus observations from a large number of case histories, they found that the H/B relationship decreased in a broadly exponential fashion with increasing B when B is plotted to a $\log_{10}$ scale. Stated another way, the relative depth to a rigid base, H/B, decreases rapidly as the foundation width increases.

Burland and Burbidge expressed this broad behavioral trend with the following empirical equation (all dimensions are in feet):

$$H = 1.4B^{0.75} \tag{5.1}$$

which is better expressed for the present discussion as:

$$\frac{H}{B} = \frac{1.4}{B^{0.25}}. \tag{5.2}$$

---

[51] The second author's surname was inadvertently and unfortunately misspelled in earlier publications by the writer. The writer regrets and apologizes for the error.

As can be seen, in the limit for large values of B the H/B ratio tends toward a value of approximately 0.3 which is substantially less than the 2-to-4 range assumed traditionally. Also, even for very narrow footings the H/B ratio is still less than the traditional value of 2. In fact, $H/B = 2$ is obtained only for an improbably small B ≈ 3 inches (75 mm). More relevant to the present discussion, a square mat foundation that is, say, 20 metres (66 ft) wide would have a depth of influence of perhaps only one-half its width, i.e. $0.5B$ (10m/33 ft), which is significantly less than the traditional rule of thumb of twice its width, i.e. $2B$ (40 m/131 ft).

In the writer's experience using the empirical relationship shown in Equations 5.1 and 5.2 in actual case history applications, one shortcoming is that it does not consider the L/B ratio explicitly in its formulation. While this is not a problem for most spread footings which tend to be square in plan view, it is not uncommon for a mat foundation to have a L/B that is substantially greater than 1. In fact, this is the case for all three case histories discussed in Chapter 6. As discussed in that chapter, one way to overcome this shortcoming, at least in an approximate manner, is to use a larger, fictitious equivalent width, $B_{equivalent}$, in these equations that is equal to the square root of the actual width, B, times the length, L. The accuracy of this approximation is addressed in Chapter 6.

Although Equations 5.1 and 5.2 are useful and provide a broad guideline, after investigation and consideration of a number of alternatives the *MT Method* (Charles 1996) was chosen by the writer as the basis of an analytical framework for estimating the depth H in Figure 5.1c. Although the Charles MT Method was developed specifically with large earthworks and landfills in mind, the writer chose to use it for SSI applications with hybrid subgrade models for several reasons:

- As originally formulated and presented, a primary outcome of the MT Method is an explicit forecast of the depth to an equivalent rigid base, H, from a settlement perspective. Consequently, the method satisfies a significant need for SSI applications, specifically, to serve as a comparative, theory-based methodology to the empirical suggestion of Burland and Burbidge (Equation 5.1).

- The overall concept behind the MT Method, i.e. that of a relatively wide loaded area so that the assumption of 1-D (vertical) stress-strain conditions within the underlying subgrade is reasonable, is sufficiently general so that it can be applied to shallow foundations of significant width such as mat foundations.

- The theoretical concepts used within the MT Method are amenable to a number of modifications, enhancements, and extensions made by the writer in order to both adapt it better to SSI applications with hybrid models as well as extend its generality in terms of subgrade conditions. The specific changes are summarized in the following items and presented in detail in Appendix D.

- The writer's modification to the method of greatest relevance to the present discussion is that an estimate of H can be produced for a much more general layered system than the subgrade assumptions made by Charles in the original MT Method. When the writer's modified method was applied to the several case histories discussed in Chapter 6, the writer found that it yielded estimates of H that were in consonance with what would be expected from results given by both Burland and Burbidge (1985) and Berardi and Lancellotta (1991) for their work more directly related to shallow foundations and was also in agreement with conclusions drawn by Mayne (2005) who used a completely different theoretical approach.

- The writer's modifications to the original MT Method allow it to explicitly consider layered systems as shown qualitatively in Figure 5.1b, something that neither the Burland and Burbidge or Berardi and Lancellotta methods are set up to deal with as their focus is more on footings with a relatively small width and relatively homogenous subgrade conditions within the assumed depth of influence.

- These same modifications made by the writer also produce the equivalent single-value modulus estimates for the assumed finite-depth elastic layer that are required by the overall process shown in Figure 5.1c. This aspect is discussed further in the following sections.

- The outcomes from the writer's modified MT Method (called the *MTH Method* when first published in Horvath (2011) and as it will be referred to hereinafter) showed overall good agreement between the forecast magnitude and distribution of settlements and measured results in the several case histories that are discussed in Chapter 6.

5.5.3 Estimation of Equivalent Single-Value Elastic Parameters

5.5.3.1 Overview

The numerical transformation of the several moduli associated with the multi-layered system shown in Figure 5.1b to the single-value moduli for the single-layer system shown in Figure 5.1c must clearly involve some type of averaging process. Typical averaging concepts that are used in geotechnical engineering when force-displacement is the primary issue as in SSI applications are:

- the arithmetic mean or 'simple' average that is based solely on the number of pieces of data without regard to the importance or influence of a given piece of data in the overall system being averaged;

- a thickness-weighted average where the thickness of the layer relative to the thickness of the overall system being averaged plays a role in determining the mathematical 'weight' or importance of a given piece of data to the overall averaging process; and

- a strain-weighted average where the strain contributed by a layer relative to the strain undergone by the total system being averaged plays a role in determining the weight given to a piece of data in the overall averaging process. Note that the strain involved is typically the normal strain parallel to the direction of loading in the specific SSI application being analyzed. In the case of mat foundations as considered here, this is the vertical normal strain that contributes to settlement.

The position adopted by the writer is that strain-weighted averaging is the preferred approach in the type of SSI applications considered in this monograph and thus the only averaging approach considered further. The following sections discuss some of the specific strategies of strain-weight averaging that the writer has investigated and, in some cases, used over the years in addition to and culminating with the writer's MTH Method noted previously.

Note that some methods mentioned in the following discussion were viable in the past but may be much less so in the present as so much has changed in terms of computational power and concomitant computational tools available to design professionals in the 40-plus years of technology that is reflected here. Nevertheless, there is at least historical value in tracing the evolution of the writer's thinking on this topic.

5.5.3.2 Fraser and Wardle (1976)

Fraser and Wardle (1976) proposed a method for determining the equivalent average Young's modulus and Poisson ratio for a mat foundation underlain by a layered subgrade system that is based on using the dimensionless strain-influence factor, *I*, from Harr's solution for the settlement beneath the corner of a perfectly flexible rectangular area with uniform applied vertical stress that is underlain by a homogenous, isotropic elastic half-space. Fraser and Wardle cited Harr (1966) as a reference but the solution can also be found in Poulos and Davis (1974) although the writer found that there are typographical errors in the equations presented in the latter publication.

Note that this strain-based weighted average only considers the vertical normal stress <u>increase</u>, $\Delta\sigma'_v$, caused by the applied load, not the <u>absolute</u> stresses that include the overburden stress. Also, only the problem geometry (length and width of the loaded area and depth to a point of interest) affect the numerical results, not the absolute value of the applied load. However, the single biggest limitation of the methodology is that a separate estimate of the depth to a rigid base must be made beforehand. This is because there must be a well-defined finite depth over which a strain-weighted average is obtained as Harr's solution is for a half-space of inherently infinite depth.

Fraser and Wardle included a chart of Harr's influence factor that could be used to perform the averaging process. However, in practice it is much easier to simply create a small computer program for this purpose. The writer did this in 1990 using FORTRAN (the program was named *EEQVH* and its use was noted in earlier publications by the writer) although nowadays it is easier to simply use spreadsheet software such as *Excel*.

5.5.3.3 Tomlinson (1986)

Tomlinson (1986) provided a snapshot of the state of knowledge for analyzing mat foundations as it existed at the time of publication. Included in the presentation was a case history example with a layered subgrade system. Given the impracticality of 3-D FE analyses at that time, the state of art for dealing with layered systems was represented by Tomlinson using two analytical methodologies:

- the *boundary-element method* (BEM) that makes use of complex integral transforms and

- the *surface-element method* (SEM) a.k.a. quasi-BEM that makes use of theory-of-linear-elasticity solutions.

Tomlinson pointed out that the BEM and SEM terms are sometimes conflated but properly refer to distinct analytical methodologies because the type of equations involved with each method are unique to that method.

The common feature of both the BEM and SEM is that the governing equations are applied only at the planar foundation-subgrade interface, with depth effects accounted for in the governing equations and not considered explicitly. In that sense, both the BEM and SEM are identical to the subgrade models that are the focus of this monograph in that the depth variable is eliminated from explicit modeling and consideration. In fact, prior publications by the writer on the subject of subgrade models have sometimes included the BEM and SEM in the discussion.

As an aside, in the context of the subgrade-modeling strategies discussed in Chapter 3, both the BEM and SEM as illustrated in Tomlinson (1986) fall into the category of what the writer defined as the geotechnical solution strategy in which the mat and subgrade are modeled and analyzed rigorously and the superstructure is not considered explicitly in terms of its interaction with the mat and subgrade. Rather, only constant loads from the superstructure are considered in the analysis. At times in the past when 3-D FE analyses were not practical even for research work, both the BEM and SEM were considered the most theoretically rigorous analytical methodologies that could be used for modeling the subgrade in SSI applications such as mat foundations and so defined the state of art at one time.

5.5.3.4 The Charles MT Method (1996) and Modifications (2006)

For the sake of completeness in the present discussion, the writer's modification (MTH Method) of the Charles MT Method that was mentioned earlier in this chapter as the writer's current preferred approach to estimating the depth to rigid base is noted here again, this time for its use in calculating strain-weighted average values of the moduli for the single-layer system depicted in Figure 5.1c.

Both the original Charles MT Method and the writer's MTH Method are discussed in detail in Appendix D. However, noted here is the fact that the benefit of both the original MT Method and the modified (MTH) version is that the strain-weight averaging is based on an estimation of the <u>absolute</u> vertical normal stresses within the subgrade, i.e. the sum of calculated overburden stresses and the estimated stress increase caused by the applied load. This is in contrast to the Fraser-Wardle approach of basing the strain-weight average on the stress <u>increase</u> only. The significance of this difference was illustrated and emphasized in Charles (1996) and is discussed in Appendix D. In addition, the MTH Method has the benefit of self-calculating the depth to rigid base and thus does not require an a priori assumption as the Fraser-Wardle methodology does.

This page intentionally left blank.

Chapter 6

Example Applications of Hybrid Subgrade Models

6.1 INTRODUCTION AND BACKGROUND

Despite the widespread use of mat foundations in practice and the publication of numerous case histories over the years (at least on a descriptive level)[52], there are surprisingly few published case histories that are usable for the desired, relatively detailed level of assessment of the hybrid subgrade models presented in this monograph. Unlike with spread footings and, especially, deep foundations where individual foundation elements can be load tested relatively easily and quickly, even to geotechnical 'failure' if desired (and still be usable afterward in many cases), to provide sufficient data for analytical comparisons an adequately documented mat foundation is a very extensive and relatively expensive effort. This is because it requires, as minimum, relatively detailed site characterization as well as precise settlement monitoring of numerous points on the mat. Furthermore, this monitoring has to continue during all phases of construction and sometimes for several years afterward when fine-grain soils are involved. Very few owners of mat-supported structures are willing to not only be supportive of such efforts, financially and otherwise, but to allow for eventual publication of measured results as they do not perceive any benefit to them. Furthermore, research-funding agencies in the U.S. at least have apparently not perceived any benefit to supporting the instrumentation and monitoring of privately owned mat-supported structures[53]. Government agencies such as the Federal Highway Administration and state departments of transportation do not, in general, build mat-supported structures as part of their mission so have no interest in supporting research efforts that will benefit mat analysis and design.

A further and significant complication with finding suitable mat foundation case histories for assessment involves the fact that mat loading is very complex. This is not only because there are numerous load points to identify and consider (not just one load as with a footing or deep-foundation element) but also because the loads themselves are not trivial to determine due to superstructure interaction effects. It has long been established that for all but the shortest buildings or simplest non-building structures that the superstructure supported by a mat will have a significant effect on both the magnitude and distribution of mat loads and resulting mat settlements due to the load redistribution that results from differential settlement of the superstructure (Burland et al. 1977, The Institution of Structural Engineers 1989, Banavalkar 1995). Very few published mat case histories provide even the most basic information concerning the superstructure to allow this important factor to be considered in even an approximate manner no less with the detail required for state-of-art analysis.

---

[52] Hemsley (2000) provides a good summary of numerous previously published case histories.

[53] The National Science Foundation turned down research proposals submitted by the writer in the late 1980s for studying subgrade models and monitoring mat foundations, with the reviewers commenting that everything that needed to be known about the analysis, design, and performance of mat foundations was already known. Keep in mind that this was still when using the Fuss-Winkler Model with a constant coefficient of subgrade reaction taken from Terzaghi's 1955 paper was the state of practice.

6.2 OVERVIEW

Despite the aforementioned challenges in finding suitable case histories for the purposes of this monograph, three were identified that were judged to be adequate to at least demonstrate the process for parameter assessment and application of the Horvath-Colasanti/Reissner (H-C/R) hybrid subgrade model in practice. Unfortunately, none of the three allowed for rigorous modeling of mat-superstructure interaction effects.

All calculated results for the three case histories that are presented in this chapter were produced in 2009 by the writer working in partnership with Mr. Regis J. Colasanti, P.E. Mr. Colasanti was responsible for performing all structural engineering calculations using the outcomes of geotechnical parameter assessment that were performed by the writer. Thus, Mr. Colasanti's contributions were essential to the outcomes report herein and his significant contributions are acknowledged with gratitude.

6.3 STRUCTURAL MODELING DETAILS OF MAT AND SUBGRADE

6.3.1 Overview

The same structural analysis software (*ANSYS* Version 11.0) and the same basic mat and subgrade models in terms of software element types were used for all three case histories presented in this chapter. Therefore, structural modeling details related to both the mat and the H-C/R subgrade model are discussed first in common.

It is recognized that in the years since these analyses were performed the *ANSYS* software has progressed through several upgrades. It is also recognized that there are many competing brands of commercially available structural analysis software that could be used instead of *ANSYS*. Therefore, in the following presentation the important aspects are not so much the specific element types used by Mr. Colasanti (which may either have been superseded or improved upon in the current version of *ANSYS*) but the technical characteristics of the element types and, especially, the reasons those characteristics were chosen for this application, details of which are presented subsequently and credited to Mr. Colasanti. It is assumed that an experienced structural engineer can use these characteristics and supporting engineering logic to find equivalent element types in other software packages of their choosing. Alternatively, if a structural engineer disagrees with the logic presented then they can make their own rational decisions as to element types to use.

Figure 6.1 is an isometric view of the overall mat plus subgrade model that was implemented into *ANSYS* 11.0 for the case history analyses presented subsequently. Note that the three different element layers that collectively comprise the mechanical component of the H-C/R subgrade model are shown in this figure separated vertically for clarity. In reality, the nodes for all layers that comprise the subgrade model (and the edge boundary condition springs as well) lie in the same plane defined by the mid-section plane of the mat elements. Stated another way, the combined mat + subgrade model has zero physical thickness in the overall structural model implemented into *ANSYS*.

Note also that if one or more planes of symmetry are used in the overall superstructure-mat-subgrade model to reduce its size (as was done for each of the case histories presented subsequently) then the special edge boundary-condition springs (denoted as K_{bc} in Figure 6.1) are not placed along the plane(s) of symmetry. These subgrade boundary-condition springs are only placed along actual edges of the mat.

Details concerning the specific *ANSYS* 11.0 element types used for all of the mat and subgrade components shown in Figure 6.1 are presented in the following sections.

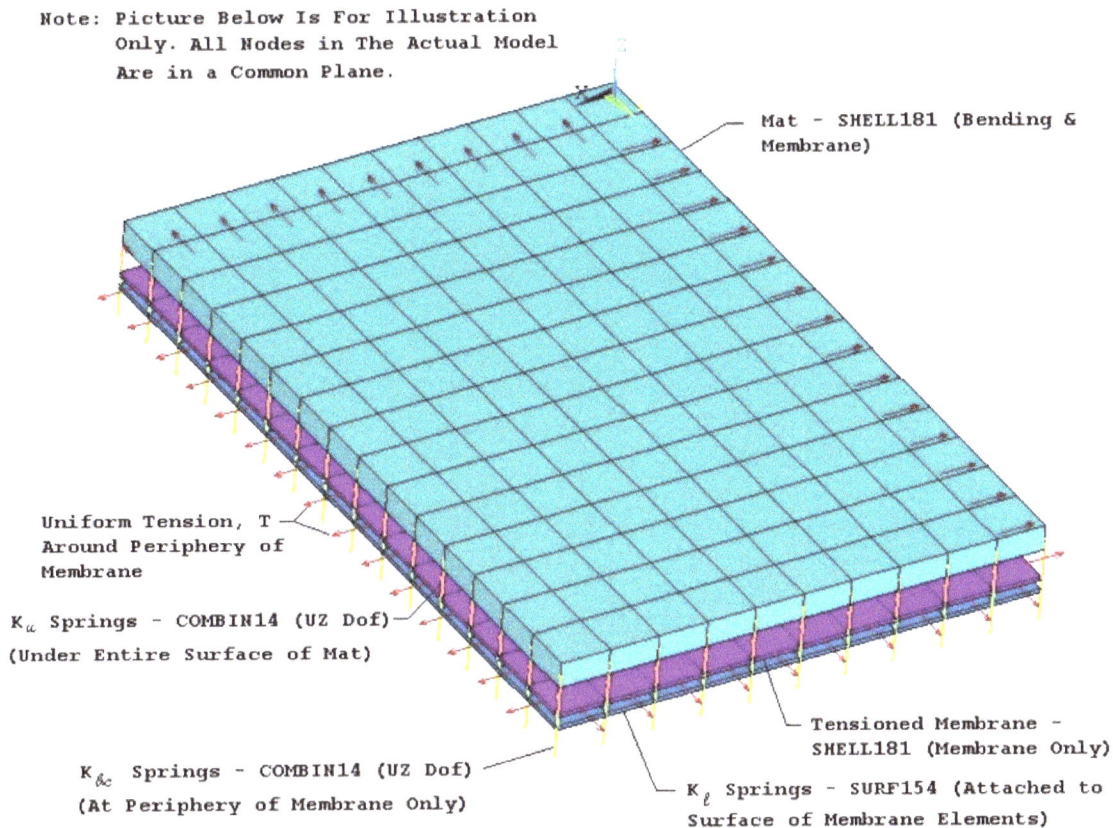

Note: Picture Below Is For Illustration
Only. All Nodes in The Actual Model
Are in a Common Plane.

Mat - SHELL181 (Bending &
Membrane)

Uniform Tension, T
Around Periphery of
Membrane

K_u Springs - COMBIN14 (UZ Dof)
(Under Entire Surface of Mat)

$K_{\&}$ Springs - COMBIN14 (UZ Dof)
(At Periphery of Membrane Only)

Tensioned Membrane -
SHELL181 (Membrane Only)

K_ℓ Springs - SURF154 (Attached to
Surface of Membrane Elements)

Figure 6.1. Implementation of the Mechanical Component of the
Horvath-Colasanti/Reissner Hybrid Subgrade Model in *ANSYS* 11.0.

6.3.2 Mat

The mat is not an inherent or explicit part of the Horvath-Colasanti/Reissner hybrid subgrade model. Therefore, the structural analyst can use whatever element type they feel is appropriate for this portion of the problem. As discussed in Chapter 2, various behavioral assumptions concerning mat behavior can be made and included in the mat model as desired by the structural analyst including:

- 'thin' vs. 'thick' plate (shear effects),

- linear vs. nonlinear (displacement/P-Δ effects),

- temporal effects on modulus (creep), and

- uncracked vs. cracked section.

As indicated in Figure 6.1, the analyses performed for the case histories reported herein used SHELL181 elements for the mat with bending and membrane action specified. This type of element includes the linear effects of transverse shear deformation which are important in the case of a relatively thick mat. Note that when using SHELL181 elements for the mat, the mesh used should consist of quadrilateral-shaped elements to the greatest extent practicable. This is because *ANSYS* strongly discouraged (as of 2009) use of triangular elements except as 'filler' elements whenever SHELL181 elements are used to model flexural behavior.

6.3.3 Subgrade Model

6.3.3.1 Overview

As discussed in Chapter 5, the mechanical component of the Horvath-Colasanti/Reissner hybrid model consists of the Horvath-Colasanti mechanical model. As shown in Figure C.1, this model consists of a 'sandwich' of upper (k_1) and lower (k_2) layers of axial springs with a deformable, constant-tension membrane (T_1) in between. Figure 6.1 shows the implementation of this mechanical model in *ANSYS* 11.0 with slightly different notation as the K_u Springs, K_l Springs, and Uniform Tension, T, respectively.

In addition, as discussed in Appendix C and illustrated in Figure C.3c, in order to minimize the size of the subgrade model and limit its horizontal extent to the plan dimensions of the mat being analyzed there is an additional perimeter row of lower springs, k_{bc}. These perimeter or edge springs, shown as the K_{bc} Springs in Figure 6.1, efficiently account for the subgrade behavior beyond the horizontal limits of the mat, at least along the exterior edges (the left and bottom edges only in Figure 6.1, the right and top edges represent planes of symmetry in the overall structural model so do not require boundary-condition springs).

As noted in Appendix C, there is a quadrant of a circle centered at each corner (only the lower-left corner in Figure 6.1) that is not accounted for with this boundary-condition modeling technique. The case history analyses presented in this chapter simply ignored the subgrade contribution from this quadrant. As discussed in Appendix C, further theoretical development is required to develop a special boundary condition model (most likely an additional, single lower spring placed only at corners) to account for the external subgrade contribution at corners.

6.3.3.2 Upper (k_1/K_u) Spring Layer

This subgrade-model component consists of a continuous bed of independent, linear, axial springs acting in the vertical direction only between the overlying mat and underlying deformable tensioned membrane. To accomplish this, COMBIN14 elements with axial stiffness in the vertical direction only were specified. Note that the actual stiffness of each of the vertical springs is a function of the area tributary to the nodes on the overlying mat and underlying deformable tensioned membrane to which that spring connects.

6.3.3.3 Deformable Tensioned Membrane (T_1/T)

This subgrade-model component consists of a continuous sheet of material with zero flexural stiffness that is under constant tension in the horizontal direction and is the sole component that produces the inherent spring coupling in the Horvath-Colasanti mechanical subgrade model. Therefore, essential to the proper performance of this subgrade model is that nonlinear (large-deflection) behavior must be specified for this membrane. Unless the deformed geometry of the membrane is considered in an analysis, the membrane does not cause the desired spring-interaction forces and the desired spring coupling will not occur.

SHELL181 elements with membrane-only behavior were specified to create this membrane. Note that element types other than SHELL181 can be used to model a tensioned membrane. For example, SHELL63 (Elastic Shell) with membrane-only behavior or SHELL41 (Membrane Shell) could be used. However, the use of these alternative element types in a large-deflection analysis come with the restriction that the element shapes must all be triangular due to possible element warping. There is no such restriction for the SHELL181 element type. Quadrilateral-shaped, membrane-only SHELL181 elements used as this element can accommodate a reasonable amount of warping. In fact, by using quadrilateral-shaped elements the mesh used for the tensioned membrane can mirror the mesh used for the mat which is also modeled using quadrilateral-shaped SHELL181 elements as discussed previously.

6.3.3.4 Lower (k₂/Kₗ) Spring Layer

This subgrade-model component consists of another continuous bed of independent, linear, axial springs acting in the vertical direction only between the overlying deformable, tensioned membrane and underlying rigid base. SURF154 elements attached to the underside of the tensioned membrane were used for this purpose. Note that COMBIN14 elements with axial stiffness in the vertical direction as used for the upper spring layer could have been used for the lower spring layer as well. However, because the lower springs are 'springs to ground', i.e. their bottom ends are restrained against vertical displacement, the use of SURF154 elements facilitates the generation of the lower springs as it allows the user to attach elements to the surface of the membrane elements and specify a property called the *foundation stiffness*, *EFS*, with dimensions of force per length per unit area. The springs generated by the SURF154 elements are then automatically internally restrained at their other (bottom in this case) end.

6.3.3.5 Edge (Perimeter) Boundary Condition (k_{bc}/K_{bc}) Springs

These springs were modeled using COMBIN14 elements with longitudinal resistance in the vertical direction only specified. Note that this is the same element type used for the upper spring layer.

6.4 CASE HISTORY: GEORGIA STATE UNIVERSITY BUILDING B

6.4.1 Overview

This 13-story dormitory building located in Atlanta, Georgia was the subject of a forensic paper by Mayne (2005). This structure was constructed in the early 1990s for initial, short-term dormitory use for the 1996 Summer Olympic Games and subsequent permanent, long-term use as an academic residence building for Georgia State University (GSU). The building settled significantly more during and after construction (approximately 10 inches/250 mm maximum) than had been forecast by the original designers (reportedly 1.8 inches/45 mm maximum).

The focus of Mayne's paper was an independent, after-the-fact reassessment of settlements using an analytical methodology based on the theory of elasticity (Mayne and Poulos 1999) and Young's moduli inferred from in-situ testing using a flat-plate dilatometer (DMT).

6.4.2 Site Geology and Geotechnical Data

The reported site geological profile consists of residual soils (fill underlain by undisturbed) transitioning to weathered bedrock and finally sound bedrock. The groundwater table was approximately 25 feet (7.5 m) below the exterior ground surface.

Mayne implies that the site conditions are relatively uniform although contours of measured mat settlements presented in Figure 7 of Mayne (2005) indicate a structure-displacement component of rigid-body tilt downward from north to south. Based on scaled data from this figure, the tilt appears to be very modest, of the order of 1:2600. This is well below tilt magnitudes that are generally considered to be problematic in practice. In any event, this tilt is indicative of either some modest variation in subsurface conditions or some variation in mat loading. Given the reported nature of the building, the former seems more likely. This is not surprising given the length of the structure relative to its width.

Mayne interpreted the DMT data for this site to be that Young's modulus was reasonably constant with depth, with a magnitude of 8500 kN/m² (180 ksf). The plot of DMT data in Figure 6 of Mayne (2005) actually indicates some variation in the interpreted Young's modulus with depth. However, given the extreme approximations in the structural loading that were necessary for back-analysis of this case history, the writer felt that performing a layered assessment along the lines of that shown in Figure 5.1 was not warranted.

Because of the specific analytical methodology used, Mayne's calculated results were very sensitive to the assumed depth to a rigid base, a situation that is conceptually identical to that faced when using hybrid subgrade models as was discussed extensively in Chapter 5 and is illustrated conceptually in Figure 5.1. Defining an actual, physical rigid base was particularly challenging for this site because of geologic conditions that are inherently gradational, a situation quite typical whenever residual soils grading to weathered bedrock then to sound bedrock are encountered as in the Piedmont region of the southeastern U.S. within which this site is located.

Because of this uncertainty, Mayne presented results for several different assumed depths to rigid base. Specifically, Mayne considered depths to rigid base within a range of 12 metres (40 ft) to 24 metres (79 ft) below foundation level of the mat. He found that 18 metres (59 ft) provided the best agreement overall between calculated and measured settlements.

6.4.3 Structural Engineering Data

Minimal information about the mat supporting this building was provided by Mayne. Only overall dimensions were provided:

- 60 feet (18 m) by 340 feet (104 m) in plan,

- 3.5 feet (1070 mm) thick, and

- founded approximately 7.5 feet (2.3 m) below grade which indicates that there was essentially no below-ground (basement) space in the building.

Mayne made no attempt to model the flexural stiffness of the mat in his analyses and assumed that it behaved in a perfectly flexible manner.

No details concerning the superstructure were provided. Based on general construction practices at the time, it is the writer's opinion that the building was likely of frame construction, possibly all PCC with a flat-plate flooring system. The building has a brick exterior wall (curtain wall or cladding) that is likely architectural, i.e. non-load-bearing, in nature. Apparently, it was visual distortion in the brick façade during the later stages of construction that was the initial indicator of unexpected differential settlements.

Mayne made no attempt to define and model actual column locations and loads, or to consider mat-superstructure interaction effects. Rather, the superstructure was modeled by Mayne as a uniform applied service-load stress of 150 kPa (3 ksf) magnitude. Note, however, that the original maximum dead plus live service loads used for design were apparently greater, approximately 200 kPa (4 ksf). Mayne gave no explanation as to why he felt that the lower stress value was more representative of actual conditions.

6.4.4 Analyses Using the Horvath-Colasanti/Reissner Hybrid Subgrade Model

6.4.4.1 Introduction

Despite the extreme simplifications regarding the structural components (mat and superstructure, individually and together) that were made by Mayne in his assessment of this case history, the results presented by Mayne were still very useful because they demonstrated that the order of magnitude of settlements actually observed could have been forecast even with these extreme structural engineering simplifications as long as a relatively accurate assessment of subgrade compressibility (Young's moduli obtained with appropriate in-situ testing combined with an appropriate analytical model) is made. Therefore, the writer felt that it would be worthwhile to see if the hybrid subgrade model approach presented in this monograph could be equally successful in its backward-forecast (backcast) of settlements.

6.4.4.2 Site Characterization

Mayne's assumption of a depth-wise constant value of Young's modulus (8500 kN/m^2 (180 ksf)) as interpreted from DMT data was used as-is so the 'artificial layered system' in this case in the context of Figure 5.1 was trivial in that it consisted of only one layer to begin with. As noted previously, the data provided by Mayne indicated some depth-wise variation in inferred modulus so that a system of artificial layers as shown in Figure 5.1b could have been created and assessed. However, given the very approximate level of detail to which structural elements and loading are known, the writer felt that such refinement in the geotechnical modeling was unwarranted analytical 'overkill'. In addition, by making the same assumptions as were made by Mayne, an arguably more interesting 'apples-to-apples' comparison of analytical outcomes was possible.

However, even with one artificial layer it was still necessary to independently estimate the depth to rigid base. This was done using the writer's MTH Method that is discussed in Appendix D. Before presenting the results obtained, it is of interest to discuss the numerical values of some of the problem parameters that Charles defined and that are used both in his original MT Method and the writer's MTH Method.

First is the effective width, b^*, of the loaded area (mat in this case) that varies between the actual width, B, and $2B$ depending on the length-to-width ratio, L/B, of the loaded area. This parameter is calculated the same way in both the MT and MTH methods. In this case, b^* = 31.1 metres (102 ft) and b^*/B = 1.70 which is indicative of a loaded area that is approaching an infinite strip behaviorally (i.e. b^*/B = 2 when L/B = ∞). This is consistent with the actual L/B = 5.7 and the fact that $L/B \geq 10$ is typically taken to be an infinite strip.

The other Charles parameter of interest is the dimensionless load intensity ratio, n, that figures prominently in the analytical algorithms of both the MT and MTH methods. While the basic definition of this parameter is the same in both versions of the method, the numerical results will, in general, be slightly different because of slight differences concerning assumptions for the parameters used to calculate n. The results were calculated for this case history only using the MTH Method and $n = 0.3$ was obtained. This is toward the low end of the range cited by Charles (from close to 100 to slightly less than 0.1) and indicative of a relatively large loaded area with a relatively low applied stress at foundation level.

The result of the MTH Method analysis for the depth to rigid base was $H = 17.7$ metres (58.0 ft) below foundation level (i.e. the mat-subgrade interface). This is a $H/B \approx 1$. This result is remarkably close to the aforementioned 18 m (59 ft) depth below foundation level that Mayne found via trial-and-error provided the best agreement overall between calculated and measured settlements.

For the sake of completeness, it is worth comparing these outcomes to results obtained using Equation 5.1 that is the empirical estimate for depth to rigid base, H, proposed by Burland and Burbidge (1985). Using the actual mat width in this equation, the calculated value of H is approximately 30 feet (9.2 m) which is $H/B \approx 0.5$ and approximately one-half of the results obtained by both the writer and Mayne. However, given the length-to-width ratio of this mat ($L/B \approx 6$), an alternative calculation using the width of an equivalent square with the same plan area, i.e. $B_{equivalent} = (B \cdot L)^{0.5} = 143$ feet (44 m), yields $H = 58$ feet (18 m) which is essentially the same as the writer's and Mayne's results.

6.4.4.3 Subgrade Model Parameters

Parameter evaluation for the mechanical component of the H-C/R Model that was implemented into *ANSYS* for the analyses performed by the writer for this case history followed the steps outlined in Appendix C. A perfectly rough interface between mat and subgrade was assumed.

Because the subgrade in this case was assumed to consist of only one artificial layer initially (corresponding to Figure 5.1b), the equivalent single-layer (Figure 5.1c) parameters were obviously the same as those for the one artificial layer. Because Young's modulus was assumed to be known, Poisson's ratio had to be assumed (0.20 in this case) and the shear modulus, G, was calculated using the usual theoretical relationship (Equation D.5). Parameter assessment for the subgrade model then followed using the equations in Appendix C and the aforementioned depth to rigid base, H, equal to 17.7 metres (58.0 ft) that was a calculated outcome of the MTH Method.

Note that both the MT and MTH methods require an assumption concerning ϕ, the Mohr-Coulomb friction angle of the subgrade geomaterials. In this case, a value of 35° was assumed by the writer.

6.4.4.4 Structural Engineering Assumptions

It is noted at the outset that the weight of the mat was neglected in the writer's analyses for this case history. The general issue of whether and how to deal with foundation-element weight in SSI applications was noted in Chapter 2 and is discussed generically in Appendix A. Mat weight was neglected for this case history because:

- groundwater was well below foundation level so the issue of uplift water pressures did not exist and

- it was assumed that the residual soils within the significant stressed depth (i.e. depth to rigid base) would compress more or less instantaneously as loads were applied due to a combination of their particle-size composition and assumed partially saturated condition so that settlement due to the weight of the mat occurred more or less contemporaneously as the mat concrete was poured. The fact that the building settled noticeably while still under construction supports this assumption that settlements occurred more or less contemporaneously with load application.

To be consistent with Mayne's assumptions, an initial analysis was performed assuming a perfectly flexible mat (i.e. 0% gross-section stiffness) with the aforementioned uniformly distributed load of 150 kPa (3 ksf). However, to provide at least a first-order assessment of the effect of non-zero mat stiffness on the calculated outcomes, a separate analysis that considered mat stiffness was also performed.

With regard to the latter, the writer's collaborator in these analyses, Mr. Colasanti, performed a rigorous set of calculations assuming (simplistically) a single-reinforced mat composed of PCC with f'_c = 4 ksi (28 MPa) and rebar only in the assumed tension zone in the lower portion of the mat. The area of steel reinforcing was sized and calculations performed in accordance with recognized U.S. design codes in effect circa 2009.

A separate structural analysis to evaluate bending moments in the mat was then performed by Mr. Colasanti[54] and cracked-section behavior based on calculated moments was checked. The conclusion was that the effective moment of inertia of the mat considering cracked-section behavior alone was 46% of the gross (uncracked) moment of inertia.

Creep and shrinkage behavior were then separately considered by Mr. Colasanti and the conclusion was that the effective mat moment of inertia considering both cracked-section and creep/shrinkage behavior was only 29% of the gross moment of inertia. This result was rounded off to 25%. Thus, the final *ANSYS* analysis performed using the H-C/R Model and assuming a mat with non-zero flexural stiffness reflected the assumption that the actual mat in the long term had only 25% of its original gross-section (uncracked) stiffness.

In both cases of assumed mat stiffness (i.e. perfectly flexible and 25% gross-section stiffness), only one-quarter of the actual mat was modeled in *ANSYS* as shown conceptually in Figure 6.1 as it was possible to take advantage of the double symmetry of the overall problem with respect to both the transverse and longitudinal directions.

No attempt was made to consider mat-superstructure interaction in the analyses performed by the writer and the effect that a 13-story (presumably) frame structure would likely have on load redistribution and the resulting distribution of settlements. Absent any information concerning column locations and other structural details, even the most basic refinement to consider mat-superstructure interaction effects did not appear to be warranted.

That having been said, the writer's MTH Method software does calculate, as an incidental outcome for information only, the dimensionless relative mat-subgrade stiffness, *K*, that was discussed in Chapter 3. This is done simply because it is a simple calculation even

---

[54] A proprietary FORTRAN program named *G183* that was developed years earlier by Mr. Colasanti was used for this purpose. This program analyzes a mat foundation supported on a layered linear-elastic system that is underlain by a rigid base using exact solutions from the theory of elasticity. As will be seen, this *G183* program was used for all three case histories presented in this monograph for various purposes.

when done manually and of interest as a 'ballpark' estimate of the relative flexibility of a mat relative to the subgrade.

As discussed in Chapter 3, over the years several researchers have proposed mathematical relationships for the K parameter, each involving the same basic parameters of the mat and subgrade. The writer calculated results using only two of the methodologies that have appeared in published works over the years. The specific results for this case history are:

- Fraser and Wardle (1976): $K = 0.77$

- Horikoshi and Randolph (1997): $K = 0.0074$.

In the overall scheme of things, these results imply behavior that is somewhat more rigid than flexible. However, these results fall within that portion of the plotted theoretical results where there is a relatively steep slope in the curve (i.e. rapid change) between perfectly flexible and perfectly rigid behavior. Stated another way, these results are within the range where small changes in K can have very different implications concerning implied relative mat-subgrade rigidity.

As noted in Chapter 3, one thing to keep in mind with all of these relationships for the K parameter is that their formulation implies that the mat will have its gross, uncracked stiffness that is unreduced by creep and shrinkage effects. As noted above, a structural assessment for the GSU mat concluded that in the long term and considering cracked-section behavior, creep, and shrinkage that the mat was likely only about one-quarter as stiff as it was initially. Because of the linear relationship between mat stiffness and K in the above relationships, this means that the above values of K are likely to decrease in magnitude to be only about one-quarter that shown in the long term, e.g. $K \approx 0.2$ for the Fraser and Wardle version of K. This implies long-term flexural behavior that is considerably more flexible and highlights the caution that should be used when using these empirical relationships for K.

That having been said, contours of the actual settlement pattern for the overall mat that are shown in Figure 7 in Mayne (2005) present a more-complex picture of relative mat flexibility. The mat appears to be relatively stiff in the transverse direction, especially near its longitudinal center. However, it is relatively flexible in its longitudinal direction and exhibits a classical sagging or dish-shaped/concave-upward settlement pattern. The four corners, especially at the south end, exhibit classical signs of relatively flexible behavior. The aforementioned component of rigid-body tilt in the longitudinal direction complicates the interpretation of settlement patterns. Nevertheless, the overall sense is that this mat, in the long term at least, exhibited moderately flexible behavior relative to the subgrade.

As an aside, this complex pattern of mat settlements for this case history points out another limitation of the K parameter. Specifically, for mats with a L/B that is substantially greater than 1 as in this case it is certainly possible for the mat to exhibit different relative flexibilities in the transverse and longitudinal directions. Thus, the one-size-fits-all conclusions inferred from calculated values of K can be misleading as they are in this case.

6.4.4.5 Results

The portrayal of vertical-profile or section views of observed settlements in Mayne's paper was somewhat atypical. Although contours of settlement for the mat in plan view were provided as is common (Figure 7 in Mayne 2005), the portrayal of settlements in vertical sections cut through the mat (Figure 9 in Mayne 2005) was atypical. Specifically, the observed settlements were plotted only for two diagonal directions between opposing corners of the mat (NW-SE and SW-NE).

In the writer's experience, it is much more common in published work to show settlement results for sections cut in directions that are oriented transverse and longitudinal to the plan dimensions of a mat. However, the plotting protocol used by Mayne does not detract from the ability to objectively compare observed and calculated results but it does make it somewhat confusing to visually comprehend the actual settlement patterns.

Before making this comparison using calculated results from the H-C/R Model, it is of value to present the results of an assessment of the relative accuracy of this subgrade model that was performed by the writer as part of the overall case history assessment. Although the relative accuracy of the H-C/R Model was explored in Horvath and Colasanti (2011a) using two simple, idealized problems, good practice dictates that assessment of this relatively new analytical methodology should continue on an ongoing basis indefinitely whenever the opportunity presents itself.

With regard to this case history, results obtained using the H-C/R Model were compared to results from two alternative analytical methodologies:

- a traditional FE analysis of a 3-D linear-elastic continuum using the same version (11.0) of *ANSYS* that was used for calculations performed using the subgrade model and

- the aforementioned proprietary *G183* program developed by Mr. Colasanti that solves the problem of a mat on a layered elastic system that incorporates closed-form solutions from the theory of elasticity.

Because each of these two analytical methodologies solves the same problem of an 'exact' 3-D linear-elastic continuum, they would be expected to yield identical results so in essence act as a check against each other in additional to being a collective basis of comparison for the H-C/R Model.

Figure 6.2 shows the calculated settlements for the two exact analytical methodologies plus the H-C/R Model along a diagonal connecting opposing corners of the GSU mat. Note that which of the two diagonals is irrelevant as the calculated results are the same for both as would be expected in this case. The numerical models used for each of the three analytical methodologies presented in this figure assumed non-zero mat stiffness (specifically, 25% of the gross-section stiffness as discussed previously) as this was judged to be more realistic than the perfectly flexible assumption.

Note that the numerical models used for all three analyses presented in this figure made use of the double symmetry of the problem analyzed so that in reality only one-quarter of the mat was actually modeled and analyzed in the respective pieces of software that were used. This means that all results are inherently symmetrical about the midpoint of the diagonal distance. However, they are shown for the full diagonal distance as this is needed subsequently for comparison to measured results that were not only different for each diagonal but were non-symmetric for each diagonal as well.

As an aside, all of these calculations were performed at an earlier stage of the overall research by the writer and Mr. Colasanti so the results are slightly different from those shown subsequently after some of the problem parameters were 'fine-tuned'. Nevertheless, because each of the three analytical methodologies depicted in Figure 6.2 solved the exact same problem and the problem parameters used differed only slightly from those used in final analyses shown subsequently, the results shown in this figure are felt to be a valid assessment of the accuracy of the H-C/R Model for the GSU case history under consideration.

Diagonal distance across mat (feet)

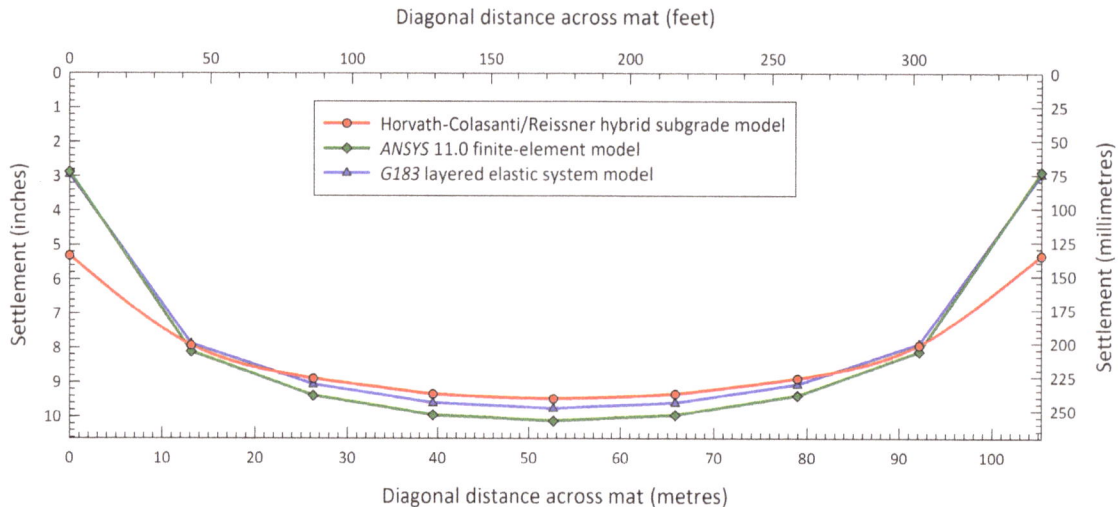

Figure 6.2. Georgia State University Building B - Subgrade Modeling Comparison.

The conclusions drawn from the results shown in Figure 6.2 are:

- The FE and layered-elastic-system results agree well along the entire diagonal distance as would be expected as both are nominally exact solutions to the problem, albeit using very different analytical methodologies.

- The H-C/R Model agrees well with the exact results except at the corners of the mat where the calculated subgrade response was noticeably 'softer' than that obtained with the exact solutions. This may be due to the way in which edge boundary conditions were modeled. As noted earlier in this chapter and discussed in detail in Appendix C, the theoretical solution for the edge/perimeter boundary condition developed by the writer for use with the H-C/R Model is theoretically correct only along continuous straight edges, not at corners. As a result, at a corner the current analytical methodology neglects a circular quadrant of subgrade response beyond the mat that is centered at a corner. Presumably, if the contribution of this missing quadrant of subgrade was included in the subgrade model then a stiffer calculated response with resulting smaller settlements would be obtained at corners. This stiffer calculated corner response would presumably bring the results obtained using the H-C/R Model into better agreement with the results obtained for a 'true' elastic continuum.

Figure 6.3 shows settlements along the aforementioned diagonals to the opposing corners of the mat. Measured results (scaled by the writer from the figures in Mayne's paper) are shown separately for the two diagonal orientations as the results for each were slightly different. On the other hand, the calculated results using the H-C/R Model (shown for both a perfectly flexible mat as Mayne assumed in his analyses and the estimated actual stiffness assuming cracked-section, creep, and shrinkage effects as described previously) are the same for either diagonal direction as noted previously. Note also that the measured results are not exactly symmetrical about midpoint of the diagonal distances as are the calculated results. This was due, at least in part, to the fact that the actual structure exhibited a modest amount of rigid-body tilt in its longitudinal direction.

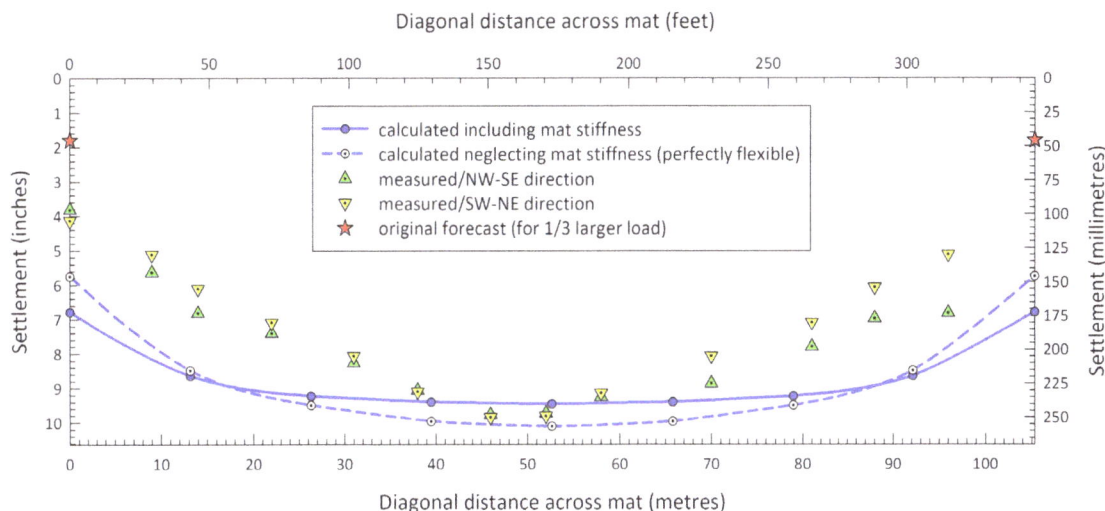

Figure 6.3. Georgia State University Building B - Settlement Comparison.

The agreement between the writer's calculated results and the measured results is considered to be very good considering all the approximations involved with the way mat loading was simulated and the fact that mat-structure interaction effects were neglected. With regard to the calculated results, there is relatively little difference between the two assumptions regarding mat flexibility. This indicates that the mat, in the long-term at least, behaved in a relatively flexible manner overall and that Mayne's assumption in his analyses that the mat was perfectly flexible was not unreasonable for the intended purpose of his calculations which were really to see if his overall analytical approach could produce 'ballpark-accurate' results which they most certainly did (as did the writer's calculated results).

Needless to say, both the measured and calculated results are substantially different from the reported maximum forecast value made during design[55]. Note that the original forecast design value was reportedly based on loads that were one-third greater than those on which both Mayne's and the writer's calculations were based. Consequently, the discrepancy between the original forecast and post-construction/forensic calculated results by both Mayne and the writer is even greater if one were to either scale-down the original forecast value or scale-up the calculated values to provide a comparison at the same load level[56].

---

[55] The reported design forecast value is shown only as individual datapoints at either end of the diagonals in Figure 6.3 as it is not known what range and/or spatial distribution of settlement estimates were made by the original designers. Mayne only reported the forecast maximum value of settlement purportedly made by the original designers.

[56] Mayne pursued this issue to some extent in his paper. The writer elected not to do so.

6.5 CASE HISTORIES: MASSACHUSETTS INSTITUTE OF TECHNOLOGY

6.5.1 Overview

DeSimone and Gould[57] (1972) presented an unusually well-documented and detailed discussion of two very similar mat-supported buildings (the Whitaker Laboratory for the Life Sciences...hereinafter referred to simply as the Whitaker Laboratory...and Chemistry Building)[58] that were constructed at MIT in Cambridge, Massachusetts approximately four years apart during the 1960s which was still very much the 'slide-rule era' in routine civil engineering practice. Their published work remains a model for a thorough, well-presented mat foundation case history paper even all these decades later.

Although these two buildings are broadly similar in terms of their geometry, design details, and construction process, their foundation loadings are geometrically quite different. As a result, the Whitaker Laboratory exhibited the common, classical sagging settlement pattern whereas the Chemistry Building displayed the less-common *hogging* (concave-downward) pattern. Consequently, they provide an interesting and informative comparison for results obtained using the H-C/R Model.

Because of the distinctly different loading and settlement patterns for these two structures, they are treated as separate case histories for the purposes of this monograph. However, because the two buildings are located relatively close to each other and share an essentially identical site geology, before discussing the two structures independently geotechnical and structural issues common to both are discussed first.

6.5.2 Site Geology and Geotechnical Data

The portion of the MIT campus within which these two buildings are located was apparently once located either within the bed of the Charles River or within the marginal wetlands along the north bank of the river. However, by the time these structures were constructed in the 1960s, the area had apparently been elevated above river level by filling some time earlier.

Of the two buildings, the Whitaker Laboratory is farther from the present river channel and parallel to it. The Chemistry Building is aligned perpendicular to the present river channel.

Holocene fill and the underlying natural Holocene organic soils (organic clay and peat) associated with the Charles River extended down to approximately 30 feet (10 m) below the ground surface that existed prior to building construction. Below that lies the well-known Boston Blue Clay (BBC), a Pleistocene marine deposit.

Both buildings had relatively deep basements and were essentially founded at the top of the BBC stratum (some localized deeper pockets of organic soils were overexcavated and replaced as part of the construction process). At the site of the Whitaker Laboratory, the thickness of the BBC stratum was uniform and approximately 70 feet (21 m) thick below foundation level. However, at the Chemistry Building site the thickness of the BBC stratum

---

[57] The writer knew both professionally on a first-hand basis during the early-1980s timeframe. DeSimone's particular foundation engineering expertise was structural design while Gould's was geotechnical.

[58] These are the building names used by DeSimone and Gould and will be used in this monograph. As of July 2018, the current names of these structures appear to be the Whitaker Building aka Building 56 and the Dreyfus Building aka Building 18 respectively.

varied somewhat, from about 70 feet (21 m) to 90 feet (27 m) below foundation level going from north to south toward the present-day channel of the river.

At both sites, the BBC stratum is underlain by a relatively thick (approximately 60 to 80 feet (18 to 24 m)) sequence of Pleistocene coarse-grain soils. These are underlain in turn by crystalline bedrock of the Cambridge Formation, an argillite of Cambrian age.

The groundwater table was relatively shallow prior to construction, within the Holocene fill and approximately 10 feet (3 m) below the ground surface that existed prior to construction and thus well above the planned foundation level for each building. The underlying Pleistocene coarse-grain stratum was a confined aquifer but no explicit piezometric information was provided.

For all practical purposes, settlements of both structures were influenced solely by compressibility of the BBC. Because the BBC is a fine-grain soil, this meant that temporal (primary consolidation) effects were significant and the long-term (drained) behavior of the BBC would govern settlement forecasts.

Data provided in DeSimone and Gould indicate that prior to construction the BBC stratum was overconsolidated relative to the current overburden stresses at its top, with the yield stress decreasing with depth until the soil became essentially normally consolidated within its lower portion. This is a very classic yield-stress pattern for fine-grain soil strata such as the BBC that are associated with Pleistocene glaciation and subsequent Holocene sea-level rise and tectonic isostasy in the northeastern U.S. In this case, the overconsolidation profile is primarily an artifact of earlier desiccation as opposed to overburden removal.

Each structure was designed to have a *compensated foundation* so that the net vertical effective stress applied at foundation level after construction was actually slightly less than the vertical effective overburden stress that existed prior to construction. This was achieved to a significant extent by designing the mat for each structure to be a *pressures slab* as opposed to a *relieved slab* so that post-construction groundwater pressures acting on the mats, once recovered and rebounded from construction, would be relatively significant...approximately 20 feet (6 m) of head...and negate a substantial portion (approximately 45% in each case) of the gross or total vertical stress imposed at foundation level by each structure. Therefore, all vertical effective stresses imposed on the BBC by these structures in the long term remained within the recompression range of the entire BBC stratum. This was likely an important design criterion for these structures in order to keep post-construction total and differential settlements within acceptable limits.

6.5.3 Structural Engineering Issues and Data

6.5.3.1 Mat Weight

The general subject of whether and how to deal with foundation-element weight in SSI applications was noted in Chapter 2 and is discussed broadly and generically in Appendix A. However, unlike for the preceding GSU case history, this issue plays a significant role for both MIT buildings. This is because the two conditions cited for the GSU building as to why mat weight was irrelevant analytically are completely the opposite for the two MIT buildings:

- the permanent, post-construction groundwater levels at the MIT sites were expected to be well above foundation levels after construction. Furthermore, both buildings were designed to have the mat function as a pressure slab, i.e. subject to 'bathtub' conditions, as opposed to as a relieved slab so uplift water pressures on the mat would exist permanently in the long term. Mat weight was a significant analytical factor in how the mat resisted and reacted to these uplift water pressures; and

- due to the fine-grain nature of the subgrade soils at the MIT site, compression could not be assumed a priori to occur more or less contemporaneously as the mat concrete was poured. As a result, there could be self-weight settlements of the mat after the concrete had cured sufficiently so that flexural stresses could develop within the mat.

Consequently, in view of the potential significance of mat weight for the MIT buildings, a detailed, site- and project-specific discussion of this issue is included here as a complement to and expansion of the generic presentation in Appendix A.

For most mat-supported structures, there are usually several distinct issues related to mat weight. Sometimes these issues can be complex and conflicting so that multiple analyses, each with a different mat-weight scenario, are required for a given structure in order to properly consider and analytically bracket all of the conditions that may exist at various times during its construction and performance life.

The specific issues that should be considered with regard to mat weight involve a complex relationship between numerous factors (most are applicable to the MIT case histories) including the:

- time it takes to excavate and dewater (as necessary) a site down to the planned foundation level;

- magnitude of subgrade heave that may occur during the excavation and dewatering process which, in turn, depends on both the compression and consolidation characteristics of the subgrade soils;

- time it takes to form and pour the mat PCC;

- time it takes for the mat PCC to transition from a fluid to a solid capable of resisting flexural stresses;

- compression and consolidation characteristics of the subgrade soils under the vertical stress imposed by the weight of the mat PCC;

- time it takes for groundwater levels to rebound and stabilize after any dewatering operations cease, especially if permanent uplift water pressures are expected to develop and remain across the bottom of the mat;

- time it takes for temporal phenomena (creep and shrinkage) to develop for the mat PCC;

- structural composition of the superstructure frame (steel versus PCC) as this will affect whether or not the frame components have material properties and concomitant flexural stiffnesses that are time independent or dependent;

- construction of the superstructure frame that will both affect and be affected by the superstructure-mat interaction;

- construction of the superstructure floors that will be affected by differential settlement;

- application of the architectural finishes (curtain wall/cladding, interior partitions) that will be affected by differential settlement;

- occupancy and/or use of the structure for its intended purpose, and whether or not the loads resulting from that use are reasonably constant or significantly variable over time; and

- cracked-section behavior of the mat that occurs as loads from the superstructure dead and live load develop over time.

As a consequence of this very complex and interactive sequence of events that is, in general, unique for a given site and structure, several distinct states can be defined as a function of time for the overall construction and post-construction process of a mat-supported structure:

- After excavation to the planned foundation level, the mat is formed and PCC poured. Depending on the size of the mat, the actual pour may extend over a period of many hours. The subgrade soils begin to compress immediately under the weight of the fluid PCC, producing settlement at the mat-subgrade interface. However, initially while the PCC is still fluid no flexural stresses are generated within the mat as a result of any differential settlements along the mat-subgrade interface induced by the weight of the still-fluid PCC. How much total settlement occurs during this stage depends on the mat dimensions as well as the compression and consolidation characteristics of the subgrade soils.

- The mat PCC begins to set and then cure. How quickly this occurs is a complex function of mat dimensions, temperature, temperature controls used (if any), and the chemistry of the PCC mix itself in terms of cement type plus additives used (if any). Depending on the characteristics of the subgrade soils, further settlement may occur during this stage even though the load applied to the subgrade is essentially constant during this time. Once the PCC begins to set, any differential settlements that occur after that point will create flexural stresses within the mat.

- The mat PCC is sufficiently cured so that construction of any below-grade walls and erection of the superstructure to be supported on the mat can begin. Note that this is typically when settlement measurement points are established on the top of the mat. This means that any excavation heave and subsequent recompression settlement under mat weight that has occurred prior to this time will be unknown unless geotechnical instrumentation has not only been placed at appropriate locations on and within the subgrade soils but has survived the construction process up to this point in time.

- As the superstructure is erected (framing, floors, and architectural walls/partitions in the case of a building) and concomitant loads are applied to the mat, additional total settlements will occur. Any differential settlements that develop will affect both the mat as well as the components of the superstructure. Note that the relationship between load application and concomitant settlement will always be site- and structure-specific depending on the compression and consolidation characteristics of the subgrade soils. In addition, if temporary dewatering has been performed as part of the excavation to foundation level then somewhere during this stage the dewatering is likely to be stopped which will allow the piezometric regime around and below the mat to begin to return to pre-construction levels, affecting the effective-stress regime of the subgrade in the process. How quickly this piezometric recovery takes will be affected by several factors, the primary one being the permeability of the subgrade soils. Changes in effective stresses within the subgrade soils as well as uplift water pressures on the bottom of the mat will complicate the progress and interpretation of any measured settlements.

- Finally, as the structure is occupied or otherwise used for its intended purpose additional total and differential settlements will occur. Again, how rapidly these progress will depend on the compression and consolidation characteristics of the subgrade soils. Depending on the specific use of the structure and the relative magnitude and time-dependent variability of live-to-dead loads on the mat, it is possible that there can be cycles of mat heave (rebound) and settlement throughout the life of the structure.

The point and conclusion of this extended discussion is that in many cases it is not clear-cut as to whether mat weight should be unilaterally included or neglected in the analysis or design of a mat-supported structure. In part, this is because the answer to this question depends on the intended use of the calculated results. For example, it should be apparent from the preceding discussion that the relevance of mat weight will vary not only as a function of the compression and consolidation characteristics of the subgrade soils but will also depend on whether the analysis is performed to:

- evaluate bending moments in the mat,

- compare forecast-to-measured settlements of the mat, or

- evaluate the superstructure for differential settlements.

Given the uncertainties involved, it is likely that in most cases analyses both without and with mat weight would or at least should be performed for a given structure and the results used to present a range of behaviors, whether for design or comparison to observations. In fact, as will be seen in the following section this is what was done for the writer's assessment of the two MIT case histories, primarily because the time-dependent behavior of the BBC subgrade soils supporting these mats greatly complicated interpretation of both calculated and measured results. However, for each structure an opinion and basis for that opinion are presented as to which case (without or with mat weight) is likely to be closer to the measured settlements for the particular mat.

6.5.3.2 Whitaker Laboratory

This building has some complex below-grade design details as a result of its being constructed adjacent to and abutting a then-pre-existing building, the Dorrance Laboratory (currently (as of July 2018) referred to as the Dorrance Building aka Building 16). As a result, the key elements of the Whitaker Laboratory are:

- eight stories above ground;

- PCC frame construction;

- superstructure is 60.0 feet (18.3 m) by 219.7 feet (67.1 m) in plan dimensions; and

- two basement levels except at one end where it abuts the Dorrance Laboratory.

The challenges presented in design of the Whitaker Laboratory revolved around the fact that the Dorrance Laboratory has only one basement level and is supported on "belled piers" that are founded at a relatively shallow depth. This resulted in the basement of the Whitaker Laboratory being stepped up from two to one basement levels for the final longitudinal column bay (a distance of approximately 20 feet/6 m) to meet the existing building basement level and not undermine it or its foundations during construction. Furthermore, that end of the Whitaker Laboratory had no superstructure foundation for that final 20± feet (6± m) so as not to stress the subgrade supporting the belled-pier foundations of the Dorrance Laboratory. As a result, that final 20± foot (6± m) portion of the Whitaker Laboratory superstructure cantilevered out from the rest of the superstructure.

The mat supporting the bulk of the Whitaker Laboratory is:

- thus overall somewhat shorter than the superstructure, approximately 200 feet (60 m) in length;

- 60.0 feet (18.3 m) wide (same as the superstructure) for the most part except on the end toward the existing Dorrance Laboratory. For approximately the final 30 feet (9 m) of mat length, the mat was widened to an unspecified dimension to accommodate the increased loads from the approximately final 20 feet (6 m) of Whitaker Laboratory superstructure that was cantilevered. Thus, the length of the Whitaker Laboratory mat that was 60.0 feet (18.3 m) wide was approximately 170 feet (52 m);

- 3.75 feet (1140 mm) thick; and

- founded 33.7 feet (10.3 m) below grade which was approximately 24 feet (7 m) below the groundwater level that existed prior to construction.

Figure 6.4 shows the assumed service loads acting on a typical transverse section of the approximately 170 feet (52 m) out of 200 feet (60 m) of mat length that was 60.0 feet (18.3 m) wide and thus considered 'typical' for this building. The three vertical forces are from the superstructure columns that were carried down through the basement to the top of the mat and the two moments are the net effect of lateral earth and water pressures acting on the exterior basement walls.

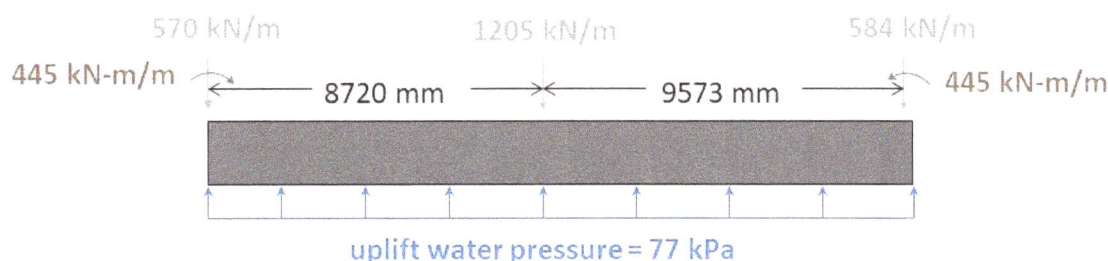

Figure 6.4. MIT Whitaker Laboratory - Mat Loads.

Only SI units are shown in this figure as these were the units used in the analyses performed by the writer that are presented subsequently. DeSimone and Gould showed essentially the same figure in their paper using Imperial units.

6.5.3.3 Chemistry Building

Because this building was isolated from pre-existing structures at the time it was constructed, the key elements of this building are overall simpler:

- five stories above ground;

- PCC frame construction;

- superstructure is 65.5 feet (20.0 m) by 279.5 feet (85.2 m) in plan dimensions;

- two basement levels throughout;

- the mat has the same plan dimensions as the superstructure and is 2.5 feet (762 mm) thick; and

- the mat is founded 30.4 feet (9.3 m) below grade which was approximately 20 feet (6 m) below the groundwater level that existed prior to construction.

This building has two features that make it significantly different than the Whitaker Laboratory from an SSI perspective:

- the mat is much thinner and thus more flexible and

- the column layout is such that all of the superstructure load is placed near the edges of the mat when viewed in a transverse cross-section. As will be seen, this produced the less-common hogging pattern of mat settlement.

The latter feature is apparent in Figure 6.5 that shows the assumed service loads acting on a typical transverse section through the mat. The four vertical forces close to the edges of the mat are from the superstructure columns that were carried down through the basement to the top of the mat. The near-center vertical force is from a column that extends only up through the basement to the ground floor so its magnitude is relatively much smaller. As with the Whitaker Laboratory, the two edge moments are the net effect of lateral earth and water pressures acting on the exterior basement walls.

Again, only SI units are shown in this figure as these were the units used in the analyses performed by the writer that are presented subsequently. DeSimone and Gould showed essentially the same figure in their paper using Imperial units.

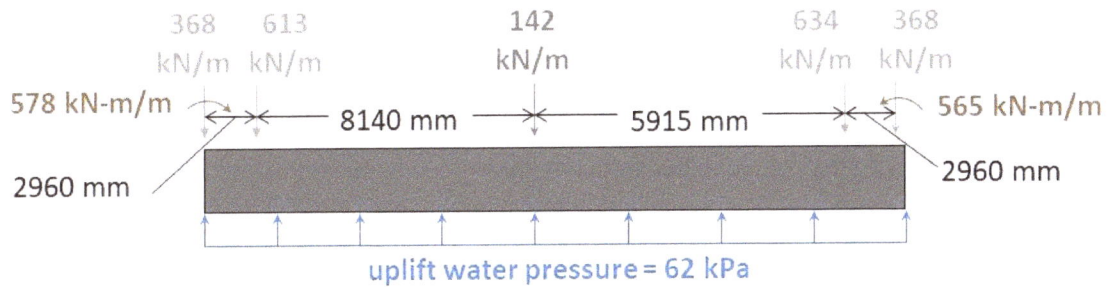

Figure 6.5. MIT Chemistry Building - Mat Loads.

6.5.3.4 Mat Structural Analyses by DeSimone and Gould

As discussed in Chapter 4, it appears that most design professionals and academicians alike have long ago forgotten Terzaghi's admonition that the sole role of subgrade models, which at that time meant only the Fuss-Winkler mechanical model with a constant coefficient of subgrade reaction, k_{FW}, with mat foundations and similar foundation elements was structural. Specifically, Terzaghi viewed the Fuss-Winkler Model as a methodology for more-accurate forecasting of local bending moments between isolated load points such as column loads that took into consideration relative mat-subgrade flexibility between load points. An outcome of this that was also noted in Chapter 4 is that Terzaghi's recommended constant values for the Fuss-Winkler coefficient of subgrade reaction historically produce values of settlement that are a fraction of one inch (a few millimetres) and generally an order of magnitude smaller than the actual settlements experienced.

It is relevant to note that DeSimone and Gould clearly understood the role that Terzaghi intended the Fuss-Winkler Model to serve. A substantial portion of DeSimone and Gould (1972) is devoted to their explaining how they used the Fuss-Winkler Model for forecasting bending moments in the mat under short-term loading (what they called an 'elastic' analysis which is in the same context as the Beam on Elastic Foundation term discussed in Chapter 4 and thus <u>not</u> to be confused with the theory of elasticity) and how they arrived at the values of the Fuss-Winkler coefficient of subgrade reaction that they used in their designs. And thus it should be no surprise that the mat settlements that they showed resulting from their 'elastic' analyses were a maximum of ¼"-½" (6-12 mm) which are but a fraction of the actual maximum settlements as discussed subsequently.

6.5.4 Analyses Using the Horvath-Colasanti/Reissner Hybrid Subgrade Model

6.5.4.1 Introduction

The two MIT case histories presented here are reasonably detailed, at least with regard to essential information concerning subsurface conditions, loading on the mats, and relatively detailed long-term settlement monitoring of the mats. Thus, they are of overall adequate detail and quality for the desired overall goal of this chapter which is to illustrate implementation of the H-C/R Model in practice.

Although critical details of the MIT building superstructures are lacking as they were with the GSU case history, the relatively modest number of stories of each MIT building suggest that superstructure-mat interaction effects were likely modest. Therefore, the fact that superstructure-mat interaction was not modeled for the analyses discussed and presented subsequently does not detract from the utility of these results for the intended purposes of this monograph.

6.5.4.2 Overall Approach to Site Characterization

The two MIT buildings share many common elements with regard to how the writer approached site characterization for the ultimate outcome of quantifying the subgrade model parameters used for the mechanical component of the H-C/R Model in *ANSYS*. Thus, it is useful to discuss these common elements before separately addressing aspects specific to each building.

To begin with, unlike with the GSU site where an obvious depth to rigid base did not exist because of the transitional nature of the residual geomaterials, the subsurface conditions at the MIT site consist of a stratigraphy that is typical of areas that have been subjected to Pleistocene glaciation, with well-defined interfaces between strata of markedly different composition and very different stiffnesses. Certainly, the top of crystalline bedrock is an unambiguous lower bound to a depth to rigid base. More than likely, the top of the coarse-grain stratum that immediately overlies bedrock would act as an even shallower upper-bound value given the fact that such soils would be expected to be significantly stiffer than the overlying normally consolidated BBC. Nevertheless, the depth to rigid base for analytical purposes was analytically and independently assessed for each building as discussed subsequently rather than relying on these visual, subjective guidelines.

The site characterization technologies available to DeSimone and Gould circa 1960 when design for the Whitaker Laboratory, which was constructed first, began were very basic and relatively primitive by today's standards. Fortunately, as it turns out, the behavior of the mat foundations for both buildings was governed entirely by compressibility of the BBC stratum which then, as now, can be adequately assessed using relatively undisturbed sampling in traditional borings followed by oedometer (one-dimensional consolidation) testing in the laboratory.

An additional factor in DeSimone and Gould's favor was that Boston is arguably the 'cradle' of modern soil mechanics in the U.S. Thus, even by circa 1960 there was a significant body of high-quality published work detailing case history experiences with the actual compressibility of the BBC by Gould himself as well as by the late Dr. Harl P. Aldrich, Jr. who had both taught at MIT and researched the settlement history of buildings on the MIT campus. This collective body of knowledge concerning the real-world compressibility of the BBC was synthesized and presented in DeSimone and Gould (1972). The net result was that for the purposes of the writer's assessment of the compressibility of the BBC the available information was more than adequate.

To begin with, the 1-D (oedometer) compressibility of the BBC was divided into two distinct zones. The first 50 feet (15 m) below foundation level of each building was overconsolidated relative to the overburden stresses that existed prior to construction and, as noted previously, the net stresses at the end of construction of the buildings were such that all stress changes would occur within the recompression range. Based on data presented in DeSimone and Gould (1972), a recompression index, C_r, = 0.02 was used for this zone of the BBC.

The remaining thickness of BBC (20 feet/6 m beneath the Whitaker Laboratory and 20 to 30 feet/6 to 9 m beneath the Chemistry Building) was normally consolidated prior to construction. However, the net stresses during and after construction remained within the recompression range so a value of the recompression index, C_r, = 0.06 was used for this zone of the BBC. Note that this is lower zone of the BBC is three times more compressible in the recompression range than the overlying zone that was overconsolidated initially.

The process of converting these basic data into an artificial, idealized layered elastic system (i.e. essentially going from Figure 5.1a to 5.1b) began by arbitrarily dividing up the BBC stratum beneath each mat into artificial layers that were 10 feet (3 m) thick with all stress calculations performed at the mid-depth of each layer. Note that there was no particular criterion for selecting this thickness other than that it seemed reasonable relative to the overall thickness of the BBC stratum being analyzed. For each layer, an operative value of the drained Young's modulus, $E_{drained}$, was calculated. The equations used to go from C_r to $E_{drained}$ are presented in Appendix D.

As can be seen from the equations in Appendix D, the key element in this process is calculating the average vertical effective stress, $\sigma'_{v\text{-}avg}$, for each artificial layer during the recompression process under the net vertical effective stress imposed at foundation level by the final structure. The following sequence of calculations was performed by the writer for this purpose for each artificial layer:

1. Calculate the vertical effective overburden stress, σ'_{vo}, prior to construction.

2. Calculate the vertical stress change, $\Delta\sigma'_v$, (negative in sign in this case) due to excavation. This needs to take into account the limited width and length of the excavation so a simple 1-D calculation is not appropriate. Any number of analytical methodologies can be used for this purpose. The writer used the simple, well-known '60° approximation' equation for this purpose:

$$\Delta\sigma'_v = p\,\frac{B \cdot L}{(B + z^*)(L + z^*)} \tag{6.1}$$

where p = the stress magnitude associated with the assumed process (excavation to foundation level in this case so negative in sign in this case), B and L are the width and length respectively of the excavation (assumed simplistically to be the same as the mat in this case), and z^* is the depth below foundation level to the mid-depth/centerline of a given artificial layer. Alternatively, some exact solution based on the theory of elasticity or even some numerical analysis could have been performed. However, the approximate calculation methodology used was felt to be adequate for the intended purpose.

3. Calculate the final vertical effective stress, $\sigma'_{v\text{-}final}$, at the end of excavation. This is the sum of the two stresses calculated in Steps 1 and 2. Note that this stress is also assumed to be the initial stress at the beginning of building construction.

4. Calculate the vertical stress change, $\Delta\sigma'_v$, (positive in sign in this case) due to the new building. The writer used Equation 6.1 for this purpose but again, any number of alternative analytical methodologies could be used. In this case, this calculation was performed both without and with mat weight but the uplift water pressure acting on the final structure was included in both cases. As an aside, as noted previously this water pressure was relatively significant for both the Whitaker Laboratory and Chemistry Building. In essence and also as noted previously, partially 'floating' each building was essential to the overall design strategy of a compensated mat foundation and keeping the final vertical effective stresses within the recompression range of the BBC, especially within the lower zone of this stratum that was normally consolidates initially.

5. Calculate the final vertical effective stress, $\sigma'_{v\text{-}final}$, for some long-term timeframe after the end of construction and recompression of the BBC is complete. This is the sum of the two stresses calculated in Steps 3 and 4.

6. The desired final outcome of the average vertical effective stress, $\sigma'_{v\text{-}avg}$, for each layer is the average of the stresses calculated in Steps 3 and 5, i.e. the beginning and end of construction respectively. Note that separate values are obtained for each artificial layer without and with mat weight.

7. From the stresses calculated in Step 6, the calculation of $E_{drained}$ for each artificial layer using the equations in Appendix D is then straightforward. Note that separate modulus values were calculated for each artificial layer for the cases without and with mat weight. For the two MIT buildings studied by the writer, the difference in moduli related to the mat-weight assumption alone was approximately ±5% for a given artificial layer.

The specific outcomes of this process for the Whitaker Laboratory and Chemistry Building are discussed individually subsequently as are the outcomes of calculating the single-value parameters for the equivalent single-layer system (i.e. going from Figure 5.1b to 5.1c) that provides the necessary inputs for the mechanical elements (k_1, k_2, T_1) of the H-C/R Model (Figure 5.1d) used in *ANSYS*.

6.5.4.3 Structural Engineering Assumptions in Common

The analyses performed using the mechanical elements of the H-C/R Model and *ANSYS* modeled the 2-D plan dimensions of the mat, albeit in an approximate manner by simply extending the 1-D loading conditions shown in Figures 6.4 and 6.5 along the full length of each mat. No attempt was made to model the basement walls and loading that actually existed at each end of each mat. Furthermore, no attempt was made to model the complex cantilever loading and localized increased mat width at the west end of the Whitaker Laboratory where it abutted the east end of the Dorrance Laboratory.

Because of these simplifications, the calculated results obtained using the H-C/R Model as implemented into *ANSYS* are felt to be most representative of conditions within the longitudinal midpoint of each mat. Fortunately, this is the portion of each mat where actual settlements were the largest in magnitude for both buildings so is the area of greatest interest. This is also the portion of each mat where DeSimone and Gould presented (in Figures 8 and 9 of their paper) their comparison of measured vs. forecast (by them) settlements. Consequently, the comparison of results presented subsequently will focus on these areas.

However, DeSimone and Gould also provided contours of measured settlements for the entire length of each mat (Figure 5 in their original paper) although they did not provide their own settlement forecasts in the longitudinal direction (if, indeed, any were made by them). Thus, some comparison of measured vs. calculated (by the writer) settlements toward and at the ends of each mat is of interest and will also be provided.

The estimated (by the writer) long-term mat stiffness was used for all calculations related to these two MIT buildings. This included an independent estimate by the writer's collaborator, Mr. Colasanti, that was made separately for each of the two buildings, of cracked-section behavior as well as creep and shrinkage as was done for the GSU case history.

Although insufficient information was provided by DeSimone and Gould to allow for explicit consideration of superstructure-mat interaction effects, the approximate rotational restraint along the edges of each mat caused by the below-grade exterior (basement) walls that are structurally connected to each mat and extend in the longitudinal direction along each edge of each mat was included in the *ANSYS* structural model used for the writer's analyses. These restraints were modeled as linear rotational springs set 1 ft (300 mm) in from each edge of each mat (the assumed midpoint of each wall). The magnitude of each rotational spring was calculated using the theoretical relationship for the unit rotational resistance of a fixed-end beam, $4EI/l$, as would be used in a matrix analysis (see Chapter 2), where EI is the assumed flexural stiffness of the basement wall and l is the height of the basement wall (approximately 30 feet (9 m) in this case).

Preliminary analyses both without and with these rotational springs were performed to assess the importance of this structural modeling detail. The relative importance was significant for both MIT buildings and the results with the rotational springs produced results much closer to the measured results. The conclusion was that even in the absence of considering superstructure-mat interaction effects attention should be paid to other structural details whenever deep basements are involved (which was not the case with the GSU building).

6.5.4.4 Whitaker Laboratory

6.5.4.4.1 Site Characterization

The process of defining artificial layers and calculating the drained Young's modulus for each layer that was outlined in detail previously yielded the results shown in Table 6.1[59] for this building. Note that the depths indicated are below foundation level to the midpoint (centerline) of a layer. The data in this table were then input into the writer's *MTH* software to produce all calculated results discussed below.

Table 6.1. MIT Whitaker Laboratory
Assumed Artificial Layers and Drained Young's Moduli.

Layer Number	Depth, z^* in metres (feet)	$E_{drained}$ in kN/m² (ksf)
1	1.53 (5)	14370 (300)
2	4.58 (15)	21076 (440)
3	7.63 (25)	27064 (565)
4	10.68 (35)	32812 (685)
5	13.73 (45)	38799 (810)
6	16.78 (55)	14610 (305)
7	19.83 (65)	16286 (340)

For the purposes of the MTH Method, the Charles effective width, b^*, in this case, is 28.8 metres (94.5 ft) and $b^*/B = 1.57$ which is indicative of a loaded area intermediate between a square ($b^*/B = 1$) and infinite strip ($b^*/B = 2$) behaviorally (the actual $L/B \approx 3.3$).

The Charles load intensity ratio, n, = 0.44. This is slightly greater than that for the GSU case history but still toward the low end of the range cited by Charles and again indicative of a relatively large loaded area with a relatively low applied stress at foundation level.

[59] The layering and moduli calculations were performed using Imperial units but input into the *MTH* software developed by the writer using SI units.

138

The result of the MTH Method analysis for the depth to rigid base was H = 19.2 metres (63.0 ft) below foundation level. This is just above the bottom of the BBC stratum (21 metres/70 ft), with a H/B slightly greater than 1.

It is again of interest to compare this outcome to results obtained using Equation 5.1 that is the empirical estimate for depth to rigid base, H, proposed by Burland and Burbidge (1985). Using the actual mat width, the calculated value of H is approximately 30 feet (9.2 m) which is H/B = 0.5 and approximately one-half of the result obtained using the MTH Method. However, given the length-to-width ratio of this mat ($L/B \approx 3.3$), an alternative calculation using the width of an equivalent square with the same plan area, i.e. $B_{equivalent} = (B \cdot L)^{0.5} = 110$ feet (33 m), yields H = 48 feet (15 m) which is still substantially less than MTH Method result in this case. This suggests that subgrade compressibility may play a role in the depth to rigid base in actual applications, something that the MTH Method inherently considers whereas the simple empirical equation of Burland and Burbidge does not.

6.5.4.4.2 Subgrade Model Parameters

As with the GSU case history, parameter evaluation for the mechanical components of the H-C/R Model that were implemented into *ANSYS* for the analyses performed by the writer for this case history followed the steps outlined in Appendix C. A perfectly rough interface between mat and subgrade was assumed.

Note that the significant difference in this case is that there is a system of multiple artificial layers (corresponding to Figure 5.1b) so the algorithm incorporated into the MTH Method was required to produce the equivalent single-layer (Figure 5.1c) parameters. Based on the assumptions that Poisson's ratio for all seven layers in Table 6.1 was equal to 0.25 and that the drained Mohr-Coulomb friction angle, ϕ, was equal to 30° for Layers 1-5 and 32° for Layers 6-7, the following single-value results were obtained:

- $E_{drained}$ = 20600 kN/m² (430 ksf) and

- G = 8230 kN/m² (172 ksf).

As discussed in Chapter 5, the MTH Method used to calculate these equivalent single-value elastic parameters is but one of several methods that can be used for this purpose. It is of interest to compare these results to those obtained using the Fraser and Wardle (1976) methodology that the writer used for many years prior to development of the MTH Method and that has a basis in the theory of elasticity.

As noted in Chapter 5, the downside of using the Fraser-Wardle Method is that an a priori assumption of the depth to rigid base, H, is required. Using the aforementioned outcome from the MTH Method analysis, H = 19.2 metres (63.0 ft), a value of $E_{drained}$ = 21520 kN/m² (450 ksf) was obtained (the corresponding value of G is trivial to calculate if desired). This is 4% greater than the value obtained using the MTH Method and considered excellent agreement given that these two analytical methodologies have nothing in common in terms of their theoretical basis.

As an aside, had the depth to rigid base from the MTH Method not been available, the writer would have assumed the depth to the bottom of the BBC stratum (21 metres (70 ft)) which is not much different. The outcome from the Fraser-Wardle Method using this value for H is $E_{drained}$ = 21020 kN/m² (440 ksf) which is 2% greater than that obtained using the MTH Method.

6.5.4.4.3 Structural Engineering Assumptions

Based on an independent analytical assessment by the writer's research collaborator, Mr. Colasanti, it was assumed that as a cumulative result of cracked-section, creep, and shrinkage behaviors that the mat for the Whitaker Laboratory had an effective, long-term value of flexural stiffness that was one-third that of the initial gross section. All analyses presented in the following section were based on this assumption.

As noted with the GSU case history, the writer's MTH Method software calculates the dimensionless relative mat-subgrade stiffness, K, that was discussed in Chapter 3. The specific results for this case history for two methods are:

- Fraser and Wardle (1976): $K = 0.31$

- Horikoshi and Randolph (1997): $K = 0.014$.

As noted for the GSU case history, it should be kept in mind that these are based on the gross, uncracked stiffness that is unreduced by creep and shrinkage effects. Thus, the effective values of these stiffnesses considering long-term effects is one-third of that shown. In addition, the comments made previously with regard to the GSU case history that a mat with a $L/B > 1$ can exhibit different relative flexibilities in the transverse and longitudinal directions applies to the Whitaker Laboratory case history as well.

6.5.4.4.4 Results

As was done for the GSU case history, before comparing measured settlements to those calculated using the H-C/R Model it is of value to present the results of an assessment of the relative accuracy of this subgrade model that was performed as part of the overall case history assessment. As noted previously, good practice dictates that assessment of this relatively new analytical methodology should continue on an ongoing basis indefinitely whenever the opportunity presents itself.

With regard to the Whitaker Laboratory, in this case the H-C/R Model was compared to one 'exact' analytical methodology, the aforementioned proprietary *G183* program that solves the problem of a mat on a layered elastic system using closed-form solutions from the theory of elasticity. Extensive prior (to 2009) use of this program by its developer, Mr. Colasanti, had shown that calculated outcomes from this program could reliably be assumed to be exact solutions to a mat supported on a system of linear-elastic layers of aggregate finite thickness. This is also reflected in the excellent comparison of calculated results between the *G183* program and a 3-D FE analysis of a layered elastic system for the GSU case history as shown in Figure 6.2.

Figure 6.6 shows the calculated settlements using the exact analytical methodology and the H-C/R Model along a typical transverse cross-section through the mat for the loading shown in Figure 6.4. Note that cases both without and with mat weight were analyzed. Note also that as with the GSU case history comparison (Figure 6.2), the results presented in Figure 6.6 were done at an earlier stage of the research before the subgrade parameters were fine-tuned. Therefore, the magnitudes of the settlements shown in Figure 6.6 differ slightly from the results of final analyses that are shown subsequently. However, the relative results between the two analytical methodologies are still valid.

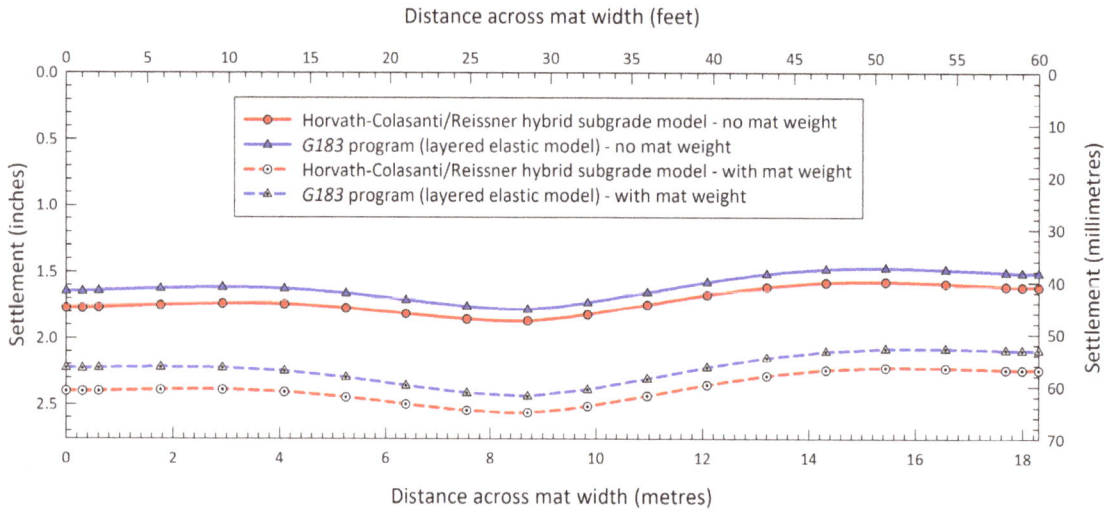

Figure 6.6. MIT Whitaker Laboratory - Subgrade Modeling Comparison.

The primary conclusions drawn from the results shown in Figure 6.6 are that the:

- H-C/R Model again does a very credible job of matching the results produced by an exact-solution methodology and

- results for this case history are very sensitive to the assumption concerning mat weight.

Figure 6.7 shows computed (for both mat-weight assumptions) vs. measured long-term (approximately six years after construction) settlements across the mat width for a region near the longitudinal center of the mat.

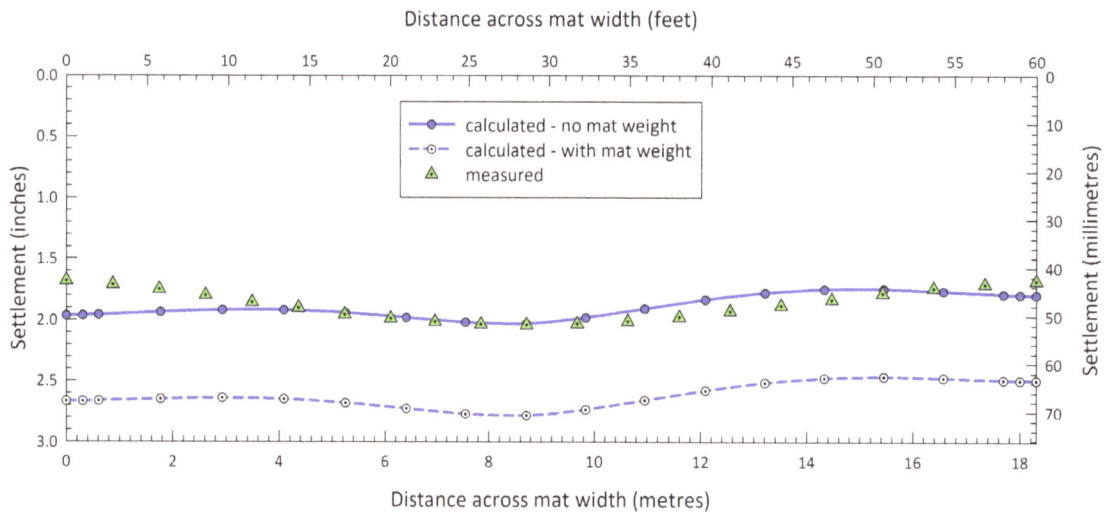

Figure 6.7. MIT Whitaker Laboratory - Settlement Comparison.

The agreement between the measured results and those calculated neglecting the weight of the mat is considered to be excellent. That the no-mat-weight assumption provided the significantly better correlation with observed settlements is consistent with the fact that field measurements presented in DeSimone and Gould (1972) suggest that the BBC had recompressed to nearly 100% primary consolidation under the mat weight before measurement points on the mat were established. This is also supported by the settlement data plotted in Figure 6 of the DeSimone and Gould paper.

As an aside, the relative significance of mat weight on the calculated settlements in this case emphasizes the point made previously that mat weight is something that should be considered carefully on each project. In this after-the-fact case, it was important to understand that even though fine-grain soils were involved, recompression under mat weight had already occurred prior to establishment of the settlement-observation points.

In any event, it is of interest to note that the maximum measured settlement (approximately 2 inches (50 mm)) was about twice the total value forecast by DeSimone and Gould[60] although their forecast pattern of settlement matched that measured very well. It is unclear why their forecast was off by approximately 100% as they presumably had access to the same oedometer data that the writer used for the results shown in Figure 6.7.

To complete the discussion of this case history, as noted previously the analyses performed by the writer for this case history modeled the longitudinal dimension of the mat, at least in an approximate manner. Thus, the calculated settlements at different longitudinal locations along the mat can be compared with the settlement contours shown in Figure 5 of the DeSimone and Gould paper.

With loadings such as that assumed for the GSU case history as well as that for the MIT Whitaker Laboratory that produce the classical sagging or dish-shaped settlement pattern, it is generally of greatest interest to compare maximum settlements at the geometric centroid (center point) of the mat to those at the corners. This is because:

- the settlement difference is theoretically greatest between these two points,

- classical theory-of-elasticity solutions typically provide results at these key locations in the form of dimensionless influence factors, and

- the relative difference provides some measure of the mat flexibility relative to the underlying subgrade.

In the case of the Whitaker Laboratory, the only relevant corner-settlement readings are for the end of the building that is farther away from the abutting Dorrance Laboratory. The measured settlement at these corners was approximately 1.2 inches (30 mm) which is about 60% of the maximum center-point settlement. These measured values compare favorably with the writer's calculated values which were approximately 1.3 inches (33 mm).

Note that using the theory of elasticity and assuming a perfectly flexible, uniformly loaded area, the percentage difference of corner to center-point settlement would always be 50% which is only slightly less than the observed 60%. Consequently, the overall behavior of the Whitaker Laboratory mat in the longitudinal direction at least was relatively flexible.

---

[60] This should not be confused with the separate value for the 'elastic' analysis performed by DeSimone and Gould for short-term, bending-moment purposes that was noted earlier in this case history presentation. The 'elastic' analysis value was only a component of the total settlement forecast made by them.

6.5.4.5 Chemistry Building

6.5.4.5.1 Site Characterization

The Chemistry Building site is slightly more complex geotechnically than the Whitaker Laboratory site due to the fact that the BBC stratum varies in thickness by approximately 20 feet (6 m) between the north and south ends of the former building. Consequently, separate site characterizations were performed for each limiting stratigraphy condition at the Chemistry Building site to see if this variation influenced the calculated results.

The same process of defining artificial layers and calculating the drained Young's modulus for each layer that was outlined in detail previously and used for the Whitaker Laboratory yielded the results shown in Table 6.2[61] for this building. Note that Layers 1 through 7 apply to both the north and south ends of the Chemistry Building while Layers 8 and 9 (shaded in gray) apply to the south end only where the BBC stratum is thicker. As with the Whitaker Laboratory, the depths indicated are below foundation level to the midpoint (centerline) of the artificial layer. The data in this table were then input into the writer's *MTH* software to produce all calculated results discussed below.

**Table 6.2. MIT Chemistry Building
Assumed Artificial Layers and Drained Young's Moduli.**

Layer Number	Depth, z^* in metres (feet)	$E_{drained}$ in kN/m^2 (ksf)
1	1.53 (5)	11496 (240)
2	4.58 (15)	17723 (370)
3	7.63 (25)	23471 (490)
4	10.68 (35)	28980 (605)
5	13.73 (45)	34249 (715)
6	16.78 (55)	13173 (275)
7	19.83 (65)	14849 (310)
8	22.88 (75)	16526 (345)
9	25.93 (85)	18202 (380)

The Charles effective width, b^*, in this case, is 32.4 metres (106 ft) and $b^*/B = 1.62$ which is indicative of a loaded area intermediate between a square ($b^*/B = 1$) and infinite strip ($b^*/B = 2$) behaviorally (the actual $L/B \approx 4.3$ which is about 30% greater than for the Whitaker Laboratory).

The Charles load intensity ratio, n, = 0.31. This is slightly lower than that for the Whitaker Laboratory and again toward the low end of the range cited by Charles. As with all the case histories presented in this chapter, it is indicative of a relatively large loaded area with a relatively low applied stress at foundation level.

The results of the MTH Method analysis for the depth to rigid base were essentially the same for both the north and south end conditions, specifically, $H = 18.7$ metres (61.3 ft)

---

[61] As with the Whitaker Laboratory, the layering and moduli calculations were performed using Imperial units but input into the *MTH* software developed by the writer using SI units.

below foundation level. This is somewhat less than that for the Whitaker Laboratory and above the bottom of the BBC stratum, with a H/B slightly less than 1.

It is again of interest to compare this outcome to results obtained using Equation 5.1 that is the empirical estimate for depth to rigid base, H, proposed by Burland and Burbidge (1985). Using the actual mat width, the calculated value of H is approximately 32 feet (10 m) which is $H/B \approx 0.5$ and approximately one-half of the results obtained using the MTH Method. However, given the length-to-width ratio of this mat ($L/B \approx 4.3$), an alternative calculation using the width of an equivalent square with the same plan area, i.e. $B_{equivalent} = (B \cdot L)^{0.5} = 135$ feet (41 m), yields $H = 56$ feet (17 m) which is only slightly less than MTH Method results in this case. Recall that for the Whitaker Laboratory the agreement was much poorer.

6.5.4.5.2 Subgrade Model Parameters

The same process of parameter evaluation for the mechanical components of the H-C/R Model that were implemented into *ANSYS* for the analyses that was used for the Whitaker Laboratory was used for the Chemistry Building. Based on the assumptions that Poisson's ratio for all nine layers in Table 6.2 was equal to 0.25 and that the drained Mohr-Coulomb friction angle, ϕ, was equal to 30° for Layers 1-5 and 32° for Layers 6-9, the following single-value results were obtained (the same for both the north and south profiles):

- $E_{drained} = 17800$ kN/m^2 (370 ksf) and

- $G = 7125$ kN/m^2 (150 ksf).

As was done for the Whitaker Laboratory, it is of interest to compare the above result for the equivalent single-layer Young's modulus obtained using the MTH Method to results obtained using the Fraser and Wardle (1976) methodology that is based on the theory of elasticity. As noted previously, the downside of using the Fraser-Wardle Method is that an a priori assumption of a depth to rigid base is required.

Considering first the results obtained when the assumed depth to rigid base, H, is that obtained using the MTH Method (18.7 m/61.3 ft in this case), a value of $E_{drained} = 18540$ kN/m^2 (390 ksf) is obtained which is 4% greater than the value obtained using the MTH Method. This is the same relative difference as for the Whitaker Laboratory and provides further support for the utility of the MTH Method.

As was opined for the Whitaker Laboratory, had the depth to rigid base from the MTH Method not been available, the writer would have assumed the depth to the bottom of the BBC stratum when using the Fraser-Wardle Method. In the case of the Chemistry Building, this would not have been substantially different than the above value of H for the north-end conditions (21 m/70 ft actual vs. 18.7 m/61.3 ft calculated) but would have been for south-end subsurface conditions (27 m/90 ft actual vs. 18.7 m/61.3 ft calculated).

To quantify this difference, analyses were performed using the Fraser-Wardle Method for both the north and south end conditions assuming the depths to rigid base were defined by the depth to the bottom of the BBC stratum. The outcomes from these analyses are $E_{drained} = 18110$ kN/m^2 (380 ksf) on the north end and $E_{drained} = 17970$ kN/m^2 (375 ksf) on the south end which are 2% and 1% greater respectively than that obtained using the MTH Method. This suggests that in at least some cases, an assessment of depth to rigid base based solely on a subjective assessment of subgrade soil conditions (in this case, a well-defined transition from normally consolidated fine-grain soil to dense coarse-grain soil) can produce reasonable results. However, the GSU case history is a reminder that for some subgrade

144

conditions...in that case a gradational morphing from residual soil to weathered rock to sound rock...such subjective assessments can be difficult to make reliably.

6.5.4.5.3 Structural Engineering Assumptions

Based on an independent analytical assessment by the writer's research collaborator, Mr. Colasanti, it was assumed that as a cumulative result of cracked-section, creep, and shrinkage behaviors that the mat for the Chemistry Building had an effective, long-term value of flexural stiffness that was one-half that of the initial gross section. All analyses presented in the following section were based on this assumption.

As noted with the two prior case histories, the writer's MTH Method software calculates the dimensionless relative mat-subgrade stiffness, K, that was discussed in Chapter 3. The specific results for this case history for two methods are:

- Fraser and Wardle (1976): $K = 0.082$

- Horikoshi and Randolph (1997): $K = 0.0021$.

As noted for the prior two case histories, it should be kept in mind that these calculated values are based on the gross, uncracked stiffness that is unreduced by creep and shrinkage effects. Thus, the effective values of these stiffnesses considering long-term effects are one-half of that shown.

When this is factored in, it is clear that these results are substantially closer to 0 (= perfectly flexible) than those for both the Whitaker Laboratory and GSU building. This is not surprising given the fact that the Chemistry Building mat is substantially thinner than that of the Whitaker Laboratory in addition to being slightly wider. As will be seen, the calculated results reflect the fact that the Chemistry Building mat was quite flexible in its forecast behavior.

6.5.4.5.4 Results

As was done for the two prior case histories, before comparing measured settlements to those calculated using the H-C/R Model it is of value to present the results of an assessment of the relative accuracy of this subgrade model that was performed as part of the overall case history assessment. As noted previously, good practice dictates that assessment of this relatively new analytical methodology should continue on an ongoing basis indefinitely whenever the opportunity presents itself.

With regard to the Chemistry Building, in this case the H-C/R Model was again compared to only one 'exact' analytical methodology, the aforementioned proprietary *G183* program that solves the problem of a mat on a layered elastic system using closed-form solutions from the theory of elasticity. Figure 6.8 shows the calculated settlements from this comparison along a typical transverse cross-section through the mat for the loading shown in Figure 6.5.

Note than an average of the north and south subsurface conditions was used in the *G183* program calculations portrayed in this figure although from a practical perspective there was virtually no difference in results obtained for the two limiting conditions. Note also that cases both without and with mat weight were analyzed.

As with the two prior case histories, the results presented in Figure 6.8 were done at an earlier stage of the research before the subgrade parameters were fine-tuned. Therefore, the magnitudes of the settlements shown in Figure 6.8 differ slightly from the results of final

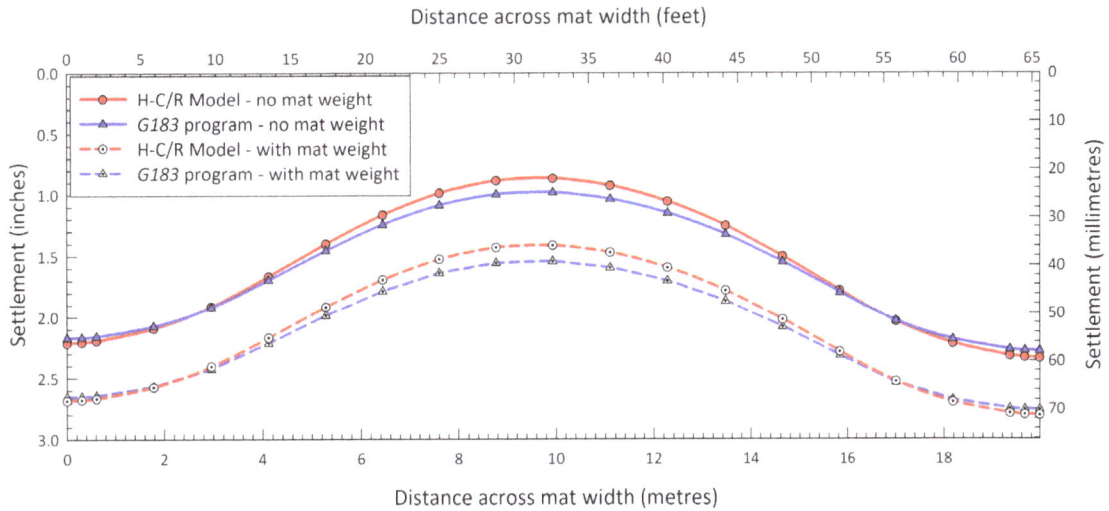

Distance across mat width (feet)

Distance across mat width (metres)

analyses that are shown subsequently. However, the relative results between the two analytical methodologies are still valid.

Figure 6.8. MIT Chemistry Building - Subgrade Modeling Comparison.

The primary conclusions drawn from the results shown in Figure 6.8 are that the:

- H-C/R Model again does a very credible job of matching the results produced by an exact-solution methodology and

- results are again very sensitive to the assumption concerning mat weight.

Figure 6.9 shows computed (for both mat-weight assumptions) vs. measured long-term (approximately two years after construction in this case) settlements across the mat width for a region near the longitudinal center of the mat (the meaning of the "extrapolated" values is discussed subsequently).

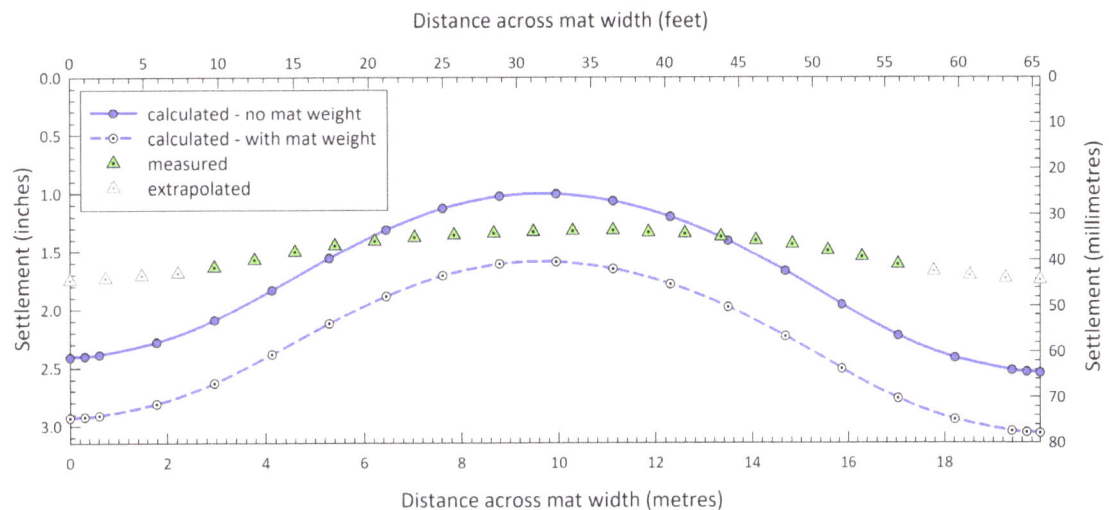

Distance across mat width (feet)

Distance across mat width (metres)

Figure 6.9. MIT Chemistry Building - Settlement Comparison.

It is obvious that the agreement between calculated vs. measured results is not very good, especially if the extrapolated results near the edges of the mat are included in the assessment, although results calculated neglecting the weight of the mat are closer to the measured results. That the no-mat-weight assumption again provided the better correlation with observed settlements is consistent with the fact that field measurements presented in DeSimone and Gould (1972) suggest that the BBC had recompressed to nearly 100% primary consolidation under the mat weight before measurement points on the mat were established. This is also supported by the settlement data plotted in Figure 6 of the DeSimone and Gould paper.

In trying to understand the source of the calculated vs. measured discrepancy in general and the fact that the measured results imply a much more rigid mat response than would be expected, it is first noted that the calculated results obtained using the H-C/R Model are believed to be 'correct' to the extent that they accurately represent outcomes based on the subsurface data and loads presented by DeSimone and Gould. This opinion was reached after noting the consonance between the H-C/R Model results and those obtained using a rigorous solution based on the theory of elasticity as shown in Figure 6.8 and discussed previously. Thus, the only logical conclusion that can be reached is that there are other reasons for the difference between calculated and measured results for this case history.

To begin with, it is relevant to note that construction of the Chemistry Building commenced about four years after that of the Whitaker Laboratory. It is possible that because of the presumably overall good experience with and performance of the Whitaker Laboratory (it was essentially completed about three years before excavation for the Chemistry Building even began) that there was less perceived need for a similarly detailed investigation for the Chemistry Building, especially because it was broadly similar in terms of its geometry and structural framing except for the difference in column layout and loading that produced a completely different settlement pattern. In particular, it is possible that the measured-settlement record for the Chemistry Building is incomplete or deficient in some way, at least relative to what was done for the Whitaker Laboratory.

Supporting this hypothesis is the fact that the overall geotechnical and structural instrumentation for the Chemistry Building, at least as presented by DeSimone and Gould, was not nearly as detailed and complete as that for the Whitaker Laboratory. For example, most of the settlement data for the Whitaker Laboratory is shown with a precision of 0.001 foot (approximately 0.01 inch/300 μm) which suggests that measurement techniques with higher-than-normal precision were utilized. Note that ordinary survey leveling of that era using manual instruments was normally assumed to have a precision of one order less, i.e. 0.01 foot (approximately ⅛ inch/3 mm) which is acceptable for many geotechnical applications.

On the other hand, the settlement data for the Chemistry Building is shown as a mixture of these two precisions. In particular, some of the data for this building carries the following note in Figure 5 in DeSimone and Gould (1972):

*"Total settlement **estimated** from earlier observations."*

with "estimated" emphasized by the writer and no elaboration as to what the "earlier observations" were. Clearly, this implies some level of uncertainty in these data.

This hypothesis is also strongly supported by two facts found in Figure 5 in DeSimone and Gould (1972) that shows the contours of measured settlements for both the Whitaker Laboratory and Chemistry Building. First, no settlement measurements were apparently made on either of the two outermost column lines of the Chemistry Building (see Figure 6.5) whereas such measurements were made for the Whitaker Laboratory (see Figure 6.4). This is why the results shown In Figure 6.9 for approximately the first 10 feet (3 m) in from each edge of the mat are shown as "extrapolated". They are not based on any actual measurements but were extrapolated by DeSimone and Gould in Figure 9 of their original paper to provide a complete profile of 'measured' settlements across the entire width of the mat for comparison to their forecast results (which again were too low, but this time by approximately 40%, not 100% as with the Whitaker Laboratory).

This is a significant omission of settlement data as the loading of the Chemistry Building is such that all of the above-ground superstructure loading is essentially transferred to the outermost 10 feet (3 m) of each edge of the mat. This is the reason for the pronounced hogging pattern of settlement forecast by the writer's analyses using both the H-C/R Model and theory-of-elasticity analysis using the *G183* program. Consequently, it is possible that the actual edge settlements were more pronounced that the extrapolations made by DeSimone and Gould.

The second fact supporting the writer's hypothesis that the settlement data presented by DeSimone and Gould for the Chemistry Building were incomplete or otherwise deficient can be found in a note contained in Figure 5 of their paper that shows the contours of measured settlements for both the Whitaker Laboratory and Chemistry Building:

"Observations for Chemistry Bldg...omit initial mat settlement."

One can only conjecture as to the magnitude and distribution of these missing data, and how they would have added to and modified that shown in Figure 6.9.

Other potential factors that may have contributed to the differences between observed and forecast settlements include the possibility that the actual mat loading differed from that presented by DeSimone and Gould and used for the writer's analyses presented here, and the effects of mat-superstructure interaction which were neglected in the present analyses except as noted previously. This latter point is thought to be quite possible given the relatively thin mat used for the Chemistry Building which means the superstructure would likely have greater influence on the overall apparent mat stiffness compared to the Whitaker Laboratory where the mat is more than three times stiffer than that of the Chemistry Building. This superstructure-interaction influence would generally be to increase the apparent mat stiffness and thus decrease differential settlements which is what was indeed observed.

To complete the discussion of the Chemistry Building case history, it is of interest to explore if the variable thickness of the BBC along the longitudinal axis of this building affected the settlements. This variation had the potential to impose a component of longitudinal, southward rigid-body tilt on the building in addition to the differential settlements that occurred in the transverse (width) direction.

The assessment of the depth to rigid base using the MTH Method indicated no difference between the two limiting subsurface stratigraphies and thus the equivalent single-value Young's modulus for the system was the same for both the north- and south-end conditions. The separate assessment using the Fraser-Wardle Method indicated an equivalent single-value modulus difference of only about 1% between the north- and south-end conditions. This suggests a very slight potential for tilting.

A review of the contoured settlements in Figure 5 of the DeSimone and Gould paper suggest that there was indeed a very slight component of rigid-body tilt toward the south, in the direction of the thicker stratum of BBC. The magnitude of this tilt was approximately ⅛" (3 mm) over the full length of the building which translates into a tilt of approximately 1:28000. Thus, for all practical purposes the rigid-body tilt of this building was nonexistent.

As a final comment on the Chemistry Building case history, because of the hogging pattern of settlement, a discussion of the relative corner to center-point settlement is not as relevant as for either the GSU or Whitaker Laboratory case histories that exhibited the much more common sagging pattern of settlement. Consequently, this issue is not addressed.

6.6 CLOSING COMMENTS RE SUBGRADE MODELING AND MODELS

In a perfect analytical world, the parameters of subgrade reaction, p, and the generic coefficient of subgrade reaction, k, that derives from it would just be calculated outcomes of secondary importance in SSI problems. In this perfect world, the topic of subgrade models would not even exist and the names of Fuss, Winkler, and many others would not be known to foundation engineers.

However, the need to decompose and/or simplify SSI problems to render them solvable in routine practice with 'ordinary' analytical resources (keeping in mind that what constitutes 'ordinary' has and is constantly evolving) has transformed subgrade reaction into an input parameter (in general and preferably as an algebraic expression and not a single value) of primary, indeed essential, importance. This fundamental paradigm shift and role reversal has created the need for subgrade models, a subject that has interested researchers since at least the earliest years of the 19th century. In the most basic of terms, a subgrade model is simply an algebraic equation that relates the subgrade reaction, p, in terms of displacement (usually settlement), w. As it turns out, the more terms in the algebraic equation that involve derivatives of p and w the more accurate the subgrade model and closer it is behaviorally to a 'true', complete soil model.

The earliest and certainly most enduring subgrade model is the Fuss-Winkler Hypothesis that famously hypothesized that the subgrade reaction at any point along a foundation-subgrade interface could be uniquely linked to only the displacement/settlement at that point via the Fuss-Winkler coefficient of subgrade reaction, k_{FW}. This constitutes what is now called a single-parameter subgrade model in that the subgrade reaction, p, is dependent on only one parameter, k_{FW}.

For the better part of century, the Fuss-Winkler Hypothesis has been recognized as being seriously flawed because it does not inherently replicate the all-important mechanism of shearing resistance (colloquially referred to as 'spring coupling') that always occurs within an actual subgrade. It is now recognized that for a subgrade model to inherently include spring coupling it must be what is called a multiple-parameter subgrade model that includes at least one additional term in the algebraic equation defining the model. This additional term involves a derivative of the displacement/settlement variable, w. The approximation of spring coupling improves as more terms involving higher derivatives of w as well as derivatives of subgrade reaction, p, are added to the algebraic equation that defines the behavior of the subgrade model.

As an aside, it is possible to indirectly include spring coupling in the Fuss-Winkler Hypothesis...sometimes this is referred to as pseudo coupling...but this requires the analyst to incorporate the spring-coupling effects into the values of the input Fuss-Winkler coefficient of subgrade reaction rather than letting the model do this inherently as in the case of multiple-parameter model. As was noted in Chapter 4, pseudo coupling amounts to knowing the correct answer beforehand so that the correct values of the input parameters can be chosen so as to yield the correct answer. While this is a seemingly impossible requirement, it has been achieved in certain SSI applications...the problem of laterally loaded deep foundations being the best known and most common...where a massive database accumulated over many decades of research and observation has allowed this seemingly impossible task to become routine reality.

In any event, as noted above, what constitutes 'ordinary' analytical practice is ever-changing. Given the current state of practice for all aspects of SSI applications...from initial site characterization to geotechnical parameter assessment to structural modeling and analysis...the broad use of the Fuss-Winkler Hypothesis is no longer intellectually defensible. This has been true now for decades although foundation engineers have been reluctant to accept this, perhaps because no clearly better <u>practical</u> alternative has existed.

This monograph in general and chapter in particular have demonstrated that using a multiple-parameter <u>hybrid</u> subgrade model that inherently incorporates the necessary spring coupling into its mathematical formulation is straightforward to use on a routine basis. The concept of a hybrid subgrade model offers the best of both worlds: a simplified-continuum component that is straightforward to quantify geotechnically and a mechanical component that is straightforward to implement into commercially available structural analysis software. The process requires geotechnical engineers in particular to actually do some engineering as opposed to looking up some tabulated value of the Fuss-Winkler coefficient of subgrade reaction in some 'cookbook' but so be it.

At the present time, the Horvath-Colasanti/Reissner Model represents the most-advanced hybrid subgrade model that has been researched and used to date. However, the comprehensive discussion of subgrade models in Chapter 4 indicates clearly that there is certainly room for improvement. This will require development of simplified-continuum subgrade models of higher order than Reissner's to complement the mechanical subgrade models of higher order that have already been developed. Subgrade boundary condition development for these higher-order hybrid models will also be required.

As final comments, although this monograph has focused on the very common and complex SSI application of mat foundations, the extension of the concepts presented herein to other categories of SSI applications is straightforward although some modification to fit a particular application will always be required.

In addition, some of the concepts developed and presented in this monograph have potential use with other analytical methodologies in geotechnical and foundation engineering. In particular, the algorithm for estimating the depth to rigid base in a layered system that the writer incorporated into the MTH Method may prove useful for other analytical applications, especially those based on the theory of elasticity. This would remove the subjectivity and guesswork that is currently required with elasticity-based methodologies. This would also provide a more-reliable procedure compared to using empirical relationships such as the one proposed by Burland and Burbidge (1985) that was shown in this chapter to have variable, unpredictable accuracy and thus variable, unpredictable reliability.

Specific examples of where an estimate of depth to rigid base that is based on the algorithm used in the MTH Method would prove useful include the Fraser-Wardle Method for performing a strain-weighted averaging of a layered elastic system to develop equivalent single-layer elastic parameters. Such parameters could be used for the simple settlement and bending moment calculations for mats that was presented in Fraser and Wardle (1976) as well as the general settlement-analysis methodology for shallow foundations and similar loaded areas that was presented by Mayne and Poulos (1999) and used in Mayne (2005) for the GSU case history presented in this chapter.

This page intentionally left blank,

Appendix A

Foundation-Element Weight Under Static Loading

A.1 INTRODUCTION

The foundation elements interacting with the ground in most SSI applications have either a horizontal or vertical orientation. One issue that arises with horizontal elements, especially those that are relatively thick such as a mat foundation, is whether or not to include their weight in calculations. For many problems, even some mat applications, this is a relatively minor issue to resolve. This is the case for one of the case histories presented in Chapter 6.

However, in some cases it is a much more consequential issue as has long been noted in the literature (e.g. Burland et al. 1977). This is also seen specifically for two of the case histories presented in Chapter 6.

Therefore, the issue of foundation-element weight in SSI applications should always be considered explicitly and dealt with rationally in order to correctly deal with those applications and projects where it may have a noticeable effect on the calculated results. The purpose of this appendix is to outline the thought process on which an application- and project-specific assessment should always be based. A specific example of how this assessment process was applied to case histories is presented in Chapter 6.

A.2 OVERVIEW

A careful assessment of the problem clearly indicates that there is no single correct answer to this question, if for no other reason that soil type (coarse- vs. fine-grain) plays a significant role due to primary-consolidation issues when fine-grain soils are involved. This will often require two sets of analyses, without and with weight, in order to evaluate the sensitivity of the final results to this assumption. This will be illustrated qualitatively using the example of a mat.

The issues regarding mat weight can be grouped into two categories:

- the effect of mat weight on mat behavior and

- the effect of mat weight on superstructure behavior.

In both cases, interaction with the subgrade is involved. As a result, there can be an effect on both of the parameters, total settlement and bending moments, that are critical for mat analysis and design.

A.3 QUALITATIVE EXAMPLES

A.3.1 Effect of Mat Weight on Mat Behavior

Mat weight causes subgrade settlement which, in turn, may induce bending moments within the mat. However, these moments will only develop after the PCC of the mat has set and achieved at least some of its strength. Thus, the time-rate at which settlement occurs is the governing factor in this case.

Settlements due to compression of coarse-grain soil or the undrained (initial) settlement component of fine-grain soil are typically presumed to occur instantaneously for analytical purposes and thus while the PCC is still fluid, thus not causing bending moments within the mat. On the other hand, the primary-consolidation settlement component of fine-grain soil develops over time and after the mat PCC has set. Thus, such settlements can create bending moments within the mat.

Note that the secondary compression (creep) component of settlement is assumed to be stress (and, therefore, weight) independent for fine-grain soils. Therefore, such settlements will occur regardless of how much the mat weighs and should be considered for their effect on mat moments as appropriate. Subgrade creep is generally best handled in SSI analyses by reducing the subgrade stiffness in a secant-modulus manner similar to that used for PCC as discussed in Chapter 2.

There is insufficient evidence at the present time as to whether or not creep settlement for coarse-grain soils is also stress independent. This settlement component is generally small in magnitude in any event and, as a result, usually neglected in routine practice.

A.3.2 Effect of Mat Weight on Superstructure Behavior

There are two opposing considerations here:

- Mat weight causes settlement as discussed in the preceding section. This may, in turn, cause the superstructure to settle as well. Again, the primary consideration is the time-rate at which settlement occurs. For coarse-grain soils and the undrained (initial) settlement component of fine-grain soils, this is generally assumed to happen instantaneously upon load application for analytical purposes. Thus, the mat will settle under its own weight before the superstructure is constructed and have no impact on superstructure settlement. On the other hand, the primary-consolidation settlement component of fine-grain soils will occur over some period of time after the mat weight is applied and, at least to some extent, after the superstructure is constructed. Therefore, in this case mat weight would have some effect on superstructure settlement[62].

- Mat weight resists post-construction uplift forces. This includes porewater pressures from groundwater that would typically be distributed over the entire width and length of the mat underside in pressure-slab design conditions as well as concentrated forces from the superstructure due to wind or seismic loading. Clearly, the beneficial effect of mat weight should be considered in these cases.

A.4 QUANTITATIVE EXAMPLES

The two case histories presented in Chapter 6 for the Massachusetts Institute of Technology (MIT) Whitaker Laboratory and Chemistry Building are excellent examples of how the generic concepts outlined in this appendix are applied to mat foundations designed to act as pressure slabs with relatively significant post-construction uplift water pressures

---

[62] The same comments made previously concerning the creep component of settlement for both fine- and coarse-grain soils apply here as well.

(approximately 20 feet/6 m of head) and constructed on a site underlain by a relatively thick deposit of fine-grain soil that completely dominates and controls settlement behavior. A very detailed presentation of the thought process that is gone through concerning mat weight as well as a comparison of calculated outcomes both without and with mat weight are presented for these two case histories in Chapter 6.

A.5 CONCLUSIONS

A decision about including the weight of the foundation element bearing on or in the ground in an SSI application should take all relevant site- and project-specific issues into account. Using the example of a mat foundation, it may be logical to both neglect and include mat weight in different loading cases of the same structure in order to properly and completely bracket the actual behavior.

For example, assume a mat on a sand subgrade. Settlement due to mat weight would reasonably be assumed to occur before the mat PCC hardens appreciably and certainly before the superstructure is erected so it would be logical to neglect mat weight for a gravity-load case. However, a load case involving post-construction uplift loads from groundwater or wind/seismic loads on the superstructure should logically include the mat weight.

Consequently, not only for this example but as a guideline for SSI analyses in general, it may be easiest to always define the foundation element as being weightless and then simulate the effect of its weight where necessary or desirable simply as an equivalent distributed applied load across the element for a particular load case.

This page intentionally left blank.

Appendix B

Conventional Method of Static Equilibrium

B.1 INTRODUCTION AND OVERVIEW

The *Conventional Method of Static Equilibrium* (CMSE) is the writer's term for an analytical methodology that is referred to in the literature authored by others using a variety of alternative terms. For example, Liao (1995) called it the *Conventional Method* and others use *Rigid Method* or some variant thereof for reasons that will become apparent subsequently.

Although not a subgrade model for SSI applications in the strict sense adopted for this monograph[63], the CMSE played an outsize role that was significant and at times dominant in SSI analyses performed in routine practice. This is especially true for mat and related foundation elements such as combined footings up until the very recent past and possibly still in the present in some cases. This is because the CMSE is the simplest and, for decades, was the most commonly used analytical method for representing subgrade behavior in SSI analyses, especially for mat foundations. The fact that it was amenable to relatively simple manual calculation if necessary or desired made it a very attractive analytical methodology even well into the computer era.

B.2 BASIC ASSUMPTIONS

The CMSE is an analytical method based on the elementary concept of rigid-body static equilibrium of forces and moments that is literally the first structural engineering subject taught to undergraduate students in civil engineering. The CMSE is always a statically determinate methodology so does not need to consider displacements and displacement compatibility in its formulation and solution as would be required in the more general matrix method of structural analysis that is discussed in Chapter 2.

As conventionally applied (what is called the 'geotechnical' interpretation for the purposes of the present discussion), the CMSE can be viewed as a logical extension of the assumptions that are made traditionally for footing foundations and the bases and footings of rigid retaining walls. This conventional geotechnical interpretation is herein explained using the example of a mat foundation:

- The vertical forces applied to the mat from the superstructure are assumed to be known beforehand from a separate structural analysis. This analysis is typically based on the assumption of no differential settlement of the superstructure. These vertical forces are summed into a single vertical resultant force and its plan-view point of application on the mat surface determined using vector statics and moment equilibrium.

- The mat is assumed to be rigid in the sense that it does not deform but only displaces (some unspecified magnitude) as a rigid body and thus will never produce differential settlements of the superstructure. In most cases, this means that the mat is assumed to

---

[63] Earlier published works by the writer did include the CMSE as a de facto subgrade model.

settle uniformly although settlement analysis is never an explicit outcome of the CMSE process. The exception to this occurs when there is an eccentricity in the resultant of the vertical forces from the superstructure with respect to the centroid of area of the mat in plan view. In this case, the mat will rotate as a rigid body (again, some unspecified magnitude of angular rotation) about one or two axes and thus settle non-uniformly but will still not produce differential settlement of the superstructure although there will be rigid-body tilt.

- The subgrade reaction, $p(x,y)$, is assumed to be completely independent of the flexural stiffness of the mat and the composition of the subgrade.

- The distribution of the subgrade reaction along the planar mat-subgrade interface is assumed to be linear (in one dimension) or planar (in two dimensions). If the mat is loaded concentrically, this means that the subgrade reaction is uniform in magnitude beneath the entire mat as is typically assumed for footing design. However, the resultant of all vertical forces applied to a mat by the superstructure will generally not coincide exactly with the centroid of area of the mat in plan view (although it should be close to minimize implied tilting of the superstructure) as it generally would for a column load on a spread footing. As a result, in this case the magnitude of the subgrade reaction is not uniform across the bottom of the foundation element but will be trapezoidal in shape on each of its four sides. The magnitude and distribution of subgrade reaction is calculated in a straightforward manner using classical structural-mechanics concepts of static equilibrium for an area loaded by both an axial force and moment (i.e. the well-known $P/A \pm Mc/I$ relationship but in this case the $\pm Mc/I$ term is applied in two horizontal dimensions), hence the name CMSE adopted many years ago by the writer.

It is relevant to note that the assumed distribution of subgrade reaction, uniform or trapezoidal, is not consistent with the actual subgrade reaction that develops beneath a relatively rigid foundation element. In reality, the subgrade reaction will have peak values around the perimeter of the foundation element (mat in this qualitative example) as illustrated in Figures 3.4 and 4.2. Therefore, the assumption of a linear distribution of subgrade reaction is idealized for analytical simplicity as it is with footings and rigid retaining walls.

B.3 STRUCTURAL ANALYSIS

B.3.1 Background and Overview

As noted previously, the function of the CMSE in SSI applications is solely to provide a basis for calculating bending moments within the foundation element (mat, combined footing, etc.) that is bearing on or in the ground. There is no provision in or adaptation to the CMSE for calculating settlements. Again, this is similar to the application of the CMSE to footings where an entirely separate analytical methodology (of which there are dozens) needs to be used to forecast settlements.

Historically and as outlined in the preceding section, there was only one interpretational approach to using the CMSE to calculate moments in a foundation element. For the purposes of the present discussion, this is referred to as the 'geotechnical' interpretation. The reason for making this distinction is that in recent decades an alternative interpretation has come to the writer's attention. This alternative is referred to herein as the 'structural' interpretation.

As will be seen in the following sections, the reason for raising this issue here is that in some cases (the only one known to the writer to date involves the base slab of a 'box' tunnel constructed using the classical cut-and-cover method) apparently different results for the same problem are obtained depending on whether the geotechnical or structural interpretational approach is used. The base slab of a box tunnel has broad similarities to a mat foundation but with some important differences as well. It is apparently these differences that has given rise to the development and use of the alternative structural interpretation.

B.3.2 Conventional 'Geotechnical' Interpretation

Because both the applied vertical forces from the superstructure and subgrade reaction magnitude and distribution are known when using the CMSE in the conventional geotechnical interpretation, the problem (at least with regard to bending moments within the mat) is rendered statically determinate as noted previously. It then becomes a straightforward structural analysis exercise to calculate those moments along with shear forces if desired.

Details concerning this conventional application of the CMSE can be found in virtually every textbook on foundation engineering (certainly older editions) as well as in professional-society committee reports such as ACI Committee 336 (1988, 1989) and earlier publications by the writer (Horvath 1979, 1983b) so are not presented here. The focus here is a critique of the CMSE methodology as well as presentation of the alternative structural interpretational methodology that, to the best of the writer's knowledge, is not addressed in textbooks and design manuals.

B.3.3 Alternative 'Structural' Interpretation

Liao (1995) noted that, at least in some applications (he specifically identified box tunnels constructed using the traditional cut-and-cover method), there is an alternative 'structural' interpretation of the CMSE that is apparently used by structural engineers in at least some applications. Using the base slab of a cut-and-cover box tunnel as an example, the specific assumptions of this structural version of the CMSE are:

- The subgrade reaction, $p(x,y)$, is always assumed to be distributed uniformly over foundation level which in this case is the horizontal contact plane between the bottom of the tunnel base slab and underlying subgrade. The magnitude of the subgrade reaction is simply the sum of all the vertical forces acting on this contact plane:
  o dead load of the tunnel structure (roof slab, exterior (side) walls, interior wall(s) if any, base slab);
  o live loads within the tunnel;
  o weight of soil cover on the tunnel roof slab;
  o surface live load (if any) above the tunnel roof slab)
  divided by the 'footprint' (plan-view) area of the base slab.

- From a structural modeling and analysis perspective, the base slab is inverted and the subgrade reaction treated as a uniformly distributed applied load acting downward. Note that this is a fundamental paradigm shift from the geotechnical interpretation discussed previously where the subgrade reaction is the resultant of applied loads from the superstructure. The forces and moments in the tunnel box are then calculated using whatever structural analysis methodology is deemed appropriate (typically the matrix method implemented into computer software nowadays). In essence, the tunnel box is modeled as a frame structure that is an inverted version of itself with a uniformly distributed load (the assumed subgrade reaction) applied across the top of the frame. Note that this represents another paradigm shift from the geotechnical interpretation in that the structure in the structural interpretation is statically indeterminate (thus requiring explicit considerations of component stiffnesses and displacements in order to calculate forces and moments) whereas in the geotechnical interpretation the problem is always statically determinate with regard to moment calculations.

In summary, the primary difference between the geotechnical and structural versions of the CMSE is that in the geotechnical version the superstructure loads are always calculated first and used as input for the geotechnical portions of the analysis whereas in the structural version the process is reversed. Also, in the geotechnical version the structural analysis of the foundation element (e.g. mat) is always rendered statically determinate. This is not the case in the structural version of the CMSE where the foundation element (e.g. tunnel base slab) is part of a frame structure that is overall statically indeterminate. As a result, it is possible that different structural designs for the same problem would be obtained from the two versions[64].

B.4 CRITIQUE

Returning now to a critique of the much more commonly used geotechnical version of the CMSE and again using the example of a mat foundation for descriptive purposes, there are a number of important issues to point out in addition to the fact that it is not a true subgrade model although it is often perceived or presented as such:

- Virtually all mats exhibit at least some flexibility. Thus, the fact that differential settlements are always zero with the CMSE is always unconservative with respect to analysis and design of the superstructure supported on the mat.

- Experience indicates that the concept of a mat foundation essentially being a large spread footing has created the unfortunate, widespread perception among many foundation and structural engineers that all that is required for design of a mat is some 'allowable bearing pressure' as with footings. As a corollary to this, there is also the perception that mat settlements are single-value as with footings. The writer has experienced this in practice countless times, including up to the relatively recent past. This simplistic way of thinking is grossly incorrect as footings and mats are quite different in this regard. For a spread footing supporting a building column, the footing dimensions in plan view are varied to match some 'allowable bearing pressure' that is based on consideration of several independent criteria:

---

[64] This difference in calculated outcomes is what brought this situation of dual interpretations of the CMSE to Liao's attention in the first place during early design stages of the well-known, world-famous Central Artery/Third Harbor Tunnel (CA/T) Project in Boston, MA, a.k.a. the Boston Big Dig (S. Liao, personal communication circa 1990). Given the size of this project, the consequences of different outcomes from analyzing the same structure in different ways had significant cost implications and was thus not a trivial or academic exercise. Note that resolution of this issue played a role in Liao's developing a unique Pseudo-Coupled Concept methodology for the Fuss-Winkler mechanical subgrade model when applied to cut-and-cover box tunnels as is discussed in Chapter 4.

o   maximum allowable settlement under service loads,

o   safety against bearing failure, and

o   code requirements.

On the other hand, the plan dimensions of a mat are generally pre-determined by the dimensions of the superstructure it will support. Thus, the designer has relatively little control over the plan dimensions of the mat and the magnitudes of total settlement, *w*, and subgrade reaction, *p*. There is only modest control over differential settlements of the mat by virtue of varying the thickness of the mat. Thus, the designer must decide whether or not the total and differential mat settlements are of acceptable magnitude for the particular superstructure in question. If not, either the superstructure design must be modified or a deep foundation or other alternative such as a piled raft used.

- Although virtually all mats exhibit some flexibility, in some cases it is modest and the actual behavior is indeed close to rigid. Nevertheless, even when a mat is relatively rigid with respect to the underlying subgrade, the assumption of a uniform or, at most, linear distribution of subgrade reaction is simply incorrect. A broad spectrum of theory (Poulos and Davis 1974); numerical modeling (Horvath 1993c, 1993d); and observation of actual foundations (Vesic and Johnson 1963, Horvath 1989b) indicates that the subgrade reaction for a 'rigid' foundation element will always be non-uniform in distribution, with well-defined maximums occurring near the edges of the foundation element as shown simplistically in Figures 3.4 and 4.2. Under such conditions, actual bending moments within the mat will always be larger than those calculated on the basis of an assumed uniform or near-uniform subgrade reaction as is done with the CMSE.

- Increasingly, engineers acknowledge that bending moments calculated using the CMSE are always incorrect, at least to some degree. However, continued use of the CMSE is still defended by some using the argument that the calculated moments, while admittedly incorrect, are at least always 'conservative' and thus, at worst, the error always results in an overly 'safe' design. However, this argument itself is simply incorrect and thus indefensible because mats always have both positive and negative moments. Experience indicates clearly that CMSE-based moments are not 'conservative' (i.e. larger in magnitude compared to an actual or analytically more-correct value) for both signs. While they are often conservative for moments of one sign, they are not for moments of the other sign. This means that a mat designed using results from a CMSE analysis will always be underdesigned in some areas from a building-code perspective.

In conclusion, the CMSE is simply unacceptable for modern SSI analyses because it does not and cannot provide estimates of total settlements. Furthermore, the CMSE always produces incorrect results on the unconservative and thus potentially unsafe side for the two primary results required for mat design:

- differential settlements and

- bending moments.

While use of the CMSE may have been defensible in the pre-computer era and when structural analysis using computers was in its infancy (especially for relatively modest structures), it no longer is and has not been for some time now[65]. In this regard, it is worth restating a relevant quotation presented in Chapter 3:

> *"Until fairly recently, there was little alternative but to proceed on the basis of greatly simplifying assumptions combined with rudimentary analysis. But although many such designs were developed with remarkable success, the limitations of this traditional approach cannot be disregarded and often are unacceptable in modern practice."*

> - John A. Hemsley, editor of *Design Applications of Raft Foundations* (2000)

---

[65] Again, with reference to the Boston Big Dig project mentioned in an earlier footnote in this appendix, this is not a trivial or academic issue as there can be significant cost implications on major projects like the Big Dig. Specifically, there were professional disagreements over use of the CMSE as opposed to 'true', more-advanced subgrade modeling for the base slabs of cut-and-cover box tunnels on this project. These disagreements were significant enough to have been mentioned in the press at the time because of costs associated with management decisions concerning use of the CMSE. As often happens, the hoary chestnuts of 'analytical conservatism' and 'safety' were trotted out for public consumption as is often done to cloak questionable technical decisions as such arguments are viewed and used... even by design professionals who should know better and be more objective...as useful shields in public debates that are right up there with 'motherhood and apple pie' as things that can neither be questioned nor debated.

Appendix C

Horvath-Colasanti/Reissner Hybrid Subgrade Model

C.1 MODEL DEVELOPMENT

C.1.1 Overview

The Horvath-Colasanti/Reissner (H-C/R) Model[66] is the hybrid subgrade model outcome of using the Horvath-Colasanti mechanical subgrade model shown in Figure C.1 to define the mechanical elements (upper spring layer + deformable tensioned membrane + lower spring layer) to be modeled using commercially available structural analysis software and the Reissner Simplified Continuum (RSC) subgrade model shown in Figure C.2 to define the mechanical-element parameters in terms of the stiffness and geometry of the actual subgrade.

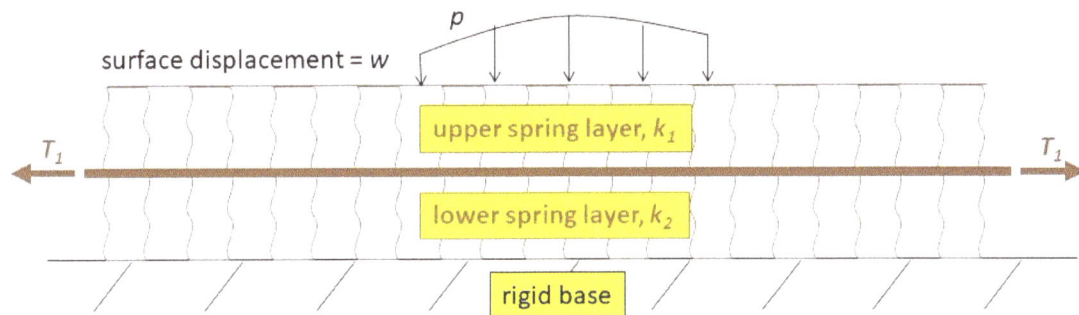

Figure C.1. Horvath-Colasanti Mechanical Subgrade Model.

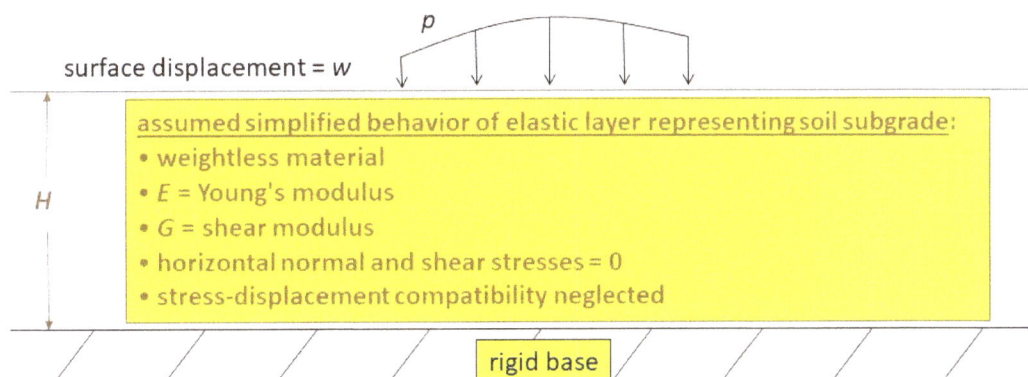

Figure C.2. Reissner Simplified Continuum Subgrade Model.

---

[66] Earlier publications by the writer such as Horvath and Colasanti (2011a) referred to this model as the Modified Kerr-Reissner (MK-R) Model.

Note that the Horvath-Colasanti Model is preferred over the mathematically equivalent Kerr Model (upper spring layer + shear-only layer + lower spring layer) in all practical applications as it easier to model a deformable tensioned membrane in commercially available structural analysis software than a shear-only layer. In fact, this was the motivation for the writer conceiving the Horvath-Colasanti Model in the first place. The writer's initial research along these lines occurred circa 1990 but it was not until approximately 20 years later that initial publication occurred in Horvath and Colasanti (2011a, 2011b) as the result of significant collaborative research concerning pragmatic structural-modeling details between the writer and Mr. Colasanti in 2009.

The governing differential equation of the Horvath-Colasanti Model was presented in Chapter 4 and is repeated here for ease of reference:

$$p - \left(\frac{T_1}{k_1 + k_2}\right)\nabla^2 p = \left(\frac{k_1 k_2}{k_1 + k_2}\right)w - \left(\frac{T_1 k_1}{k_1 + k_2}\right)\nabla^2 w \qquad \textbf{(C.1)}$$

with all parameters defined in Figure C.1.

Details of how the three mechanical elements (k_1, k_2, T_1) that define the behavior of the Horvath-Colasanti Model were implemented into a specific brand and version of software (*ANSYS* Version 11.0) circa 2009 were presented originally in Colasanti and Horvath (2011). A graphic from that paper that illustrates the specific subgrade model implemented into *ANSYS* 11.0 is reproduced as Figure 6.1 as well as on the back cover of this monograph.

It is recognized that there are numerous competing brands of commercially available structural analysis software, each with their own element types. However, the overall modeling results described in Colasanti and Horvath (2011) are expected to be achievable with any state-of-art software capable of performing a nonlinear analysis. Therefore, the detailed reasoning behind the types of *ANSYS* 11.0 elements used to produce the calculated results shown in Colasanti and Horvath (2011) that was presented in that paper is repeated in Chapter 6 of this monograph. This is intended to provide generic reference and guidance for structural analysts selecting element types in the specific software that they are using.

As for the numerical values of the k_1, k_2, and T_1 parameters in the Horvath-Colasanti Model, they are obtained by correlation with the RSC Model coefficients[67]. This is the heart of the hybridization process that links geotechnically measurable properties of a subgrade with abstract mechanical parameters.

As discussed in Chapter 4, the governing equation of the RSC varies slightly depending on the presence of a foundation element (beam or plate) on the subgrade surface and, if a foundation element is present, whether the contact surface between the underside of the element and the subgrade surface is assumed to be either perfectly smooth or perfectly rough. Consequently, it is necessary to develop two sets of parameter correlations between the Horvath-Colasanti mechanical model and the RSC depending on the particular case assumed in a given project-specific application.

C.1.2 No Foundation/Perfectly Smooth Foundation

The governing differential equation for the case of no foundation element or one that has a perfectly smooth interface with the subgrade was presented in Chapter 4 and is repeated here for ease of reference:

---

[67] Assessment of the RSC Model coefficients was originally addressed in Horvath (2011), and an expanded and updated treatment of this subject is presented in Chapters 5 and 6 of this monograph.

$$p - \left(\frac{GH^2}{12E}\right)\nabla^2 p = \left(\frac{E}{H}\right)w - \left(\frac{GH}{3}\right)\nabla^2 w . \tag{C.2}$$

Note that this is Reissner's original solution (Reissner 1958) with some rearrangement of the coefficients and viscoelastic effects neglected.

Correlating the coefficients between Equations C.1 and C.2 produces three equations with three unknowns and yields the following desired final results:

$$k_1 = \frac{4E}{H}, \tag{C.3a}$$

$$k_2 = \frac{4E}{3H}, \tag{C.3b}$$

$$T_1 = \frac{4GH}{9} . \tag{C.3c}$$

As an aside, it is of interest to note that the upper spring layer is substantially stiffer than the lower spring layer which implies that most of the 1-D compression of the overall subgrade system emanates from the lower spring layer. Because only the lower spring layer determines the relative vertical displacement of the deformable tensioned membrane, this means that the lower spring layer completely determines the magnitude of 'spring coupling' (vertical shearing resistance) that occurs.

The relative spring stiffnesses in this subgrade model are also of interest with regard to the subgrade boundary conditions discussed later in this appendix as all of the subgrade compression that occurs beyond the limits of the loaded area/foundation element occurs in the lower spring layer.

C.1.3 Perfectly Rough Foundation

The governing differential equation for the case of a foundation element that has a perfectly rough interface with the subgrade was presented in Chapter 4 and is repeated here for ease of reference:

$$p - \left(\frac{GH^2}{12E}\right)\nabla^2 p = \left(\frac{E}{H}\right)w - \left[\left(\frac{GH}{2}\right)\left(\frac{2}{3} - \frac{t}{2H}\right)\right]\nabla^2 w . \tag{C.4}$$

where t is the thickness of the foundation element. Derivation of this case was an original contribution of the writer (Horvath 1979).

Correlating the coefficients between Equations C.1 and C.4 produces three equations with three unknowns and yields the following desired final results:

$$k_1 = \frac{E}{H}\left(\frac{4H - 3t}{H}\right), \tag{C.5a}$$

$$k_2 = \frac{E}{3H}\left(\frac{4H - 3t}{H - t}\right), \tag{C.5b}$$

$$T_1 = \frac{GH}{12}\left[\left(\frac{4H - 3t}{H}\right) + \left(\frac{4H - 3t}{3H - 3t}\right)\right].$$ (C.5c)

As expected, if $t = 0$ in Equations C.5a-c then Equations C.3a-c are recovered.

C.2 SUBGRADE BOUNDARY CONDITION DEVELOPMENT

C.2.1 Introduction

As noted throughout Chapter 4, whenever a multiple-parameter subgrade model is used, whether it is one based on mechanical elements as in the case of the Horvath-Colasanti Model or a simplified elastic continuum as in the case of the RSC, the behavior of the subgrade beyond the edge of the subgrade reaction, p, must be considered. This is because, by inspection for any multiple-parameter model, the subgrade settlement, w, is not zero even where the subgrade reaction, p, is zero. This is due to the spring coupling (in mechanical models) and load spreading via vertical shear (in simplified-continuum models) that mimics actual subgrade behavior. Thus, only the single-parameter Fuss-Winkler Hypothesis and the subgrade models derived from it or mathematically equivalent to it (such as the writer's Fuss-Winkler Type Simplified Continuum) are inherently immune from this boundary-condition requirement.

For structural modeling and analytical efficiency, it is always desirable to not explicitly model the unloaded portion of the subgrade but rather replace its effects by some subgrade boundary condition applied along the edge of the loaded area or foundation element. Developing the necessary boundary conditions requires additional analytical work beyond developing the basic governing differential equation of a subgrade model. Note that this additional analytical work:

- is not generic and is always particular to a given model,

- does not appear to have been done by every researcher for every subgrade model that has been proposed over the years, and

- is not unique for a given model (as evidenced by the different opinions of A. D. Kerr/Rhines vs. Reissner as discussed at length in Chapter 4) and will produce very different results depending on the assumptions made so should always be clearly noted in any use of a model.

C.2.2 Background and Overview

Colasanti and Horvath (2011) presented a detailed derivation of the basic subgrade boundary condition used with the Horvath-Colasanti Model (it could also be used for the mathematically equivalent Kerr Model with minor conceptual and notational changes) and, therefore, in applications of the H-C/R hybrid model in practice. This derivation is repeated in this appendix for ease of reference.

Figure C.3a is a generic elevation view through the edge of a mat of thickness t supported on a Horvath-Colasanti mechanical model subgrade that represents the mechanical component of the Horvath-Colasanti/Reissner hybrid subgrade model and is thus a visualization of what is modeled in commercially available structural analysis software. A straight edge of unspecified length in the direction perpendicular to this figure represents the basic subgrade boundary condition for this problem.

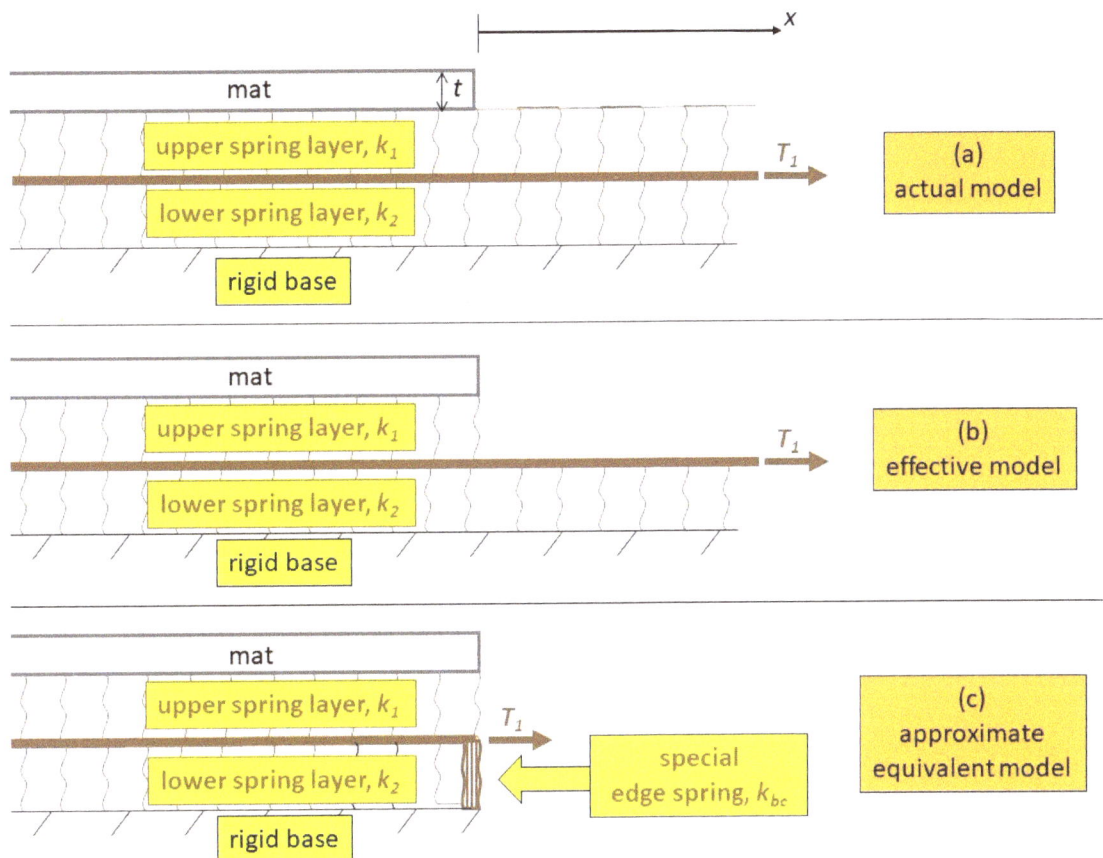

Figure C.3. Basic Subgrade Boundary Condition for Horvath-Colasanti Model.

To eliminate the need to explicitly model the subgrade beyond the edge of the subgrade reaction, p (which is coincidental with the edge of the mat in this case), the first step is to deal with the four boundary conditions that must be considered at the edge of the mat. Two are the usual structural ones for the mat itself and are easily handled. Specifically, either the settlement of or shear within the mat must be zero, and either the rotation of or moment within the mat must be zero. For most mats, the edge shears and moments will be zero so that the edge of the mat is free to settle and rotate without restraint.

The other two boundary conditions involve the subgrade, specifically the first derivatives of both the subgrade reaction, p, and surface settlement, w. To resolve these boundary conditions, continuity of the vertical shear forces within the subgrade from the inside to the outside of the mat is assumed as was suggested originally by Reissner (1958) for the mathematically equivalent RSC. This allows the edge of the mat to 'punch' into the subgrade.

As discussed in Chapter 4, others who have worked extensively with multiple-parameter subgrade models (primarily A. D. Kerr and his protégé, Rhines) have chosen the other possible alternative which is continuity of surface settlement, w, at the edge of the mat.

This assumption generates a smooth settlement profile but with a concentrated shear force at the edge of the mat.

Both assumptions (continuity of shear per Reissner vs. continuity of settlement per A. D. Kerr and Rhines) were investigated extensively in the past by the writer using a propriety FORTRAN-based computer code named *SSIH* that was purpose-written by the writer beginning in 1990 (the most recent revision is Horvath (2009a)) specifically for research purposes to evaluate different subgrade models as well as other effects that are relevant to mat foundation behavior. This program is based on the matrix method and couples various subgrade models with various boundary-condition options for the foundation element. The foundation element is also coupled to a simple superstructure frame to allow mat-superstructure interaction to be studied in at least a primitive manner and allows for analytical variations such as incremental loading and simulation of progressive cracked-section behavior of the mat using the well-known Branson equation/formula. The calculated results using various subgrade and structural modeling assumptions were presented in the past in Horvath (1993c, 1993d).

Based on other, unpublished results obtained using the *SSIH* program, it was found that only Reissner's continuity-of-shear assumption produces results that are both intuitively pleasing and representative of actual subgrade behavior, namely, a gradual buildup of subgrade reaction, p, toward the edge of the mat is generated that replicates the expected generic behavior as shown in Figures 3.4 and 4.2. The continuity-of-settlement assumption promoted by A. D. Kerr and Rhines was found to produce a uniform distribution of subgrade reaction across the entire mat combined with a concentrated vertical-upward force at the edge of the mat. This is judged to be less representative of actual subgrade reactions and SSI behavior.

Using the continuity-of-shear assumption, the subgrade boundary condition reduces to the problem of determining the magnitude and sense (upward or downward) of the net vertical shear force that must be applied to the edge of the mat by the subgrade to replicate the resistance to settlement produced by the subgrade beyond the edge of the mat. The goal is to reduce this shear force to a mechanical element or elements placed continuously along the edge of the mat so that the element(s) can be implemented in commercially available structural analysis software.

To start the process, the following are assumed:

- Settlements occur only in one horizontal dimension perpendicular to the edge of the mat (depicted in Figure C.3a as the x-axis, with the origin of this axis being the edge of the mat) so partial derivatives can be replaced by total derivatives. As discussed subsequently, this results in an element of approximation in places where the edge of the mat changes direction in plan view such as at corners.

- The subgrade reaction, p, and all of its derivatives are equal to zero beyond the edge of the mat.

- The upper spring layer beyond the edge of the mat does not contribute to the subgrade resistance so this spring layer can be eliminated from consideration as shown in Figure C.3b.

The net result of these assumptions is that the subgrade beyond the edge of the mat for $x > 0$ effectively reduces to that shown in Figure C.3b which is physically simpler than the original Horvath-Colasanti mechanical subgrade model beneath the mat.

The resulting, effective mechanical subgrade model beyond the edge of the mat, which consists of a deformed/deformable tensioned membrane over an axial-spring layer, is no longer the Horvath-Colasanti Model of Order H^3 but the simpler Filonenko-Borodich Model of Order H^2. Of historical note is that the latter mechanical subgrade model, which was presented and discussed in Chapter 4, is believed to have been the first advanced (i.e. multiple-parameter) mechanical subgrade model that was proposed in the English-language published literature, circa 1940.

C.2.3 Derivation of Edge Boundary Condition

The governing differential equation of the Filonenko-Borodich Model is given in Chapter 4 but is repeated here in a slightly modified form[68] for ease of reference:

$$p = k_2 w - T_1 \nabla^2 w .$$ (C.6)

Beyond the edge of a mat, $p = 0$ and settlement is only in the x-axis direction so Equation C.6 simplifies to (with rearrangement of the constant coefficients):

$$\frac{d^2 w}{dx^2} - \frac{k_2}{T_1} w = 0 .$$ (C.7)

The generic solution to Equation C.7 is:

$$w(x) = C_1 e^{\left(\sqrt{\frac{k_2}{T_1}}\right)x} + C_2 e^{\left(-\sqrt{\frac{k_2}{T_1}}\right)x} .$$ (C.8)

Noting that:

➤ as $x \to \infty$, $w(x) \to 0$, $\therefore C_1 = 0$

➤ @ $x = 0$, $w(x) = w(0) \stackrel{\text{def}}{=} w_0$, $\therefore C_2 = w_0$

produces:

$$w(x) = w_0 e^{\left(-\sqrt{\frac{k_2}{T_1}}\right)x}$$ (C.9)

and

$$\frac{dw(x)}{dx} = \theta(x) = \left(-\sqrt{\frac{k_2}{T_1}}\right) w_0 e^{\left(-\sqrt{\frac{k_2}{T_1}}\right)x} = \left(-\sqrt{\frac{k_2}{T_1}}\right) w(x)$$ (C.10)

where $\theta(x)$ is the slope of the deformed/deformable tensioned membrane.

---

[68] The subscript of the spring stiffness in this equation has been changed from that shown in Chapter 4 from '1' to '2' to be consistent with Figure C.3b and the fact that this equation as used in this context represents the subgrade behavior beyond the edge of mat with only the lower (k_2) spring layer present.

168

At the edge of the mat ($x = 0$), $\theta(x) = \theta(0) \stackrel{\text{def}}{=} \theta_0$. The vertical component of the force produced by the deformed/deformable tensioned membrane at $x = 0$ is $T{\cdot}\sin\theta_0 \approx T{\cdot}\theta_0$ assuming θ_0 is a relatively small angle.

θ_0 is obtained by evaluating Equation C.10 at $x = 0$ and yields a value of $\left(-\sqrt{\frac{k_2}{T_1}}\right)w_0$.

Note that the negative sign here simply confirms mathematically that this force acts upward as would be expected by inspection. Therefore, the net vertical force produced at the edge of the mat by the collective subgrade effects beyond the edge of the mat has an upward orientation and absolute-value magnitude of:

$$T_1\theta_0 = T_1\left|\left(\sqrt{\frac{k_2}{T_1}}\right)w_0\right|. \tag{C.11}$$

The desired final result is achieved by dividing the resultant force given by Equation C.11 by w_0 to produce an equivalent axial spring of stiffness k_{bc} that is defined by:

$$k_{bc} = \frac{T_1\left|\left(\sqrt{\frac{k_2}{T_1}}\right)w_0\right|}{w_0} = T_1\sqrt{\frac{k_2}{T_1}} = \sqrt{T_1^2\left(\sqrt{\frac{k_2}{T_1}}\right)} = \sqrt{T_1^2\frac{k_2}{T_1}} = \sqrt{k_2 T_1}. \tag{C.12}$$

Note that k_{bc} is the same as the springs labeled K_{bc} in Figure 6.1 and the back cover of this monograph. The notation has been changed in this monograph for reasons that will become clear shortly.

Thus, the net subgrade effects of the Horvath-Colasanti mechanical model beyond the edge of the mat are easily replaced and replicated by an additional continuous line of independent axial springs placed along the edge of the mat between the deformed/deformable tensioned membrane and rigid base, i.e. at the same level as the lower spring layer. This is illustrated conceptually in Figure C.3c.

Within the context of the hybrid H-C/R Model as would be used in practice, the magnitude of the spring stiffness k_{bc} per unit length of mat edge is obtained by evaluating the abstract parameters k_2 and T_1 in the final result shown in Equation C.12 using either Equations C.3b and C.3c for a perfectly smooth mat-subgrade interface or Equations C.5b and C.5c for a perfectly rough interface. An interesting result of doing this is that the depth to rigid base, H, cancels out so that k_{bc} is a function only of the subgrade stiffnesses E and G.

C.2.4 Limitations of Edge Boundary Condition

Note that dealing with the edge boundary condition in the manner outlined above is rigorously correct along a straight edge of unlimited length but is somewhat approximate overall in actual applications as effects at corners are not dealt with rigorously and completely. This is because the above derivation assumed that all subgrade effects beyond the edge of the mat only occur in a direction perpendicular to that edge. Thus, at corners there is a radial zone (90° in most cases) of subgrade beyond the corner of the mat and between the two straight-edge segments that meet at that corner that is essentially neglected as it falls in between the two edge boundary conditions that meet at that corner. How to deal with this omission is discussed in the following sections.

C.2.5 Special Case of a Corner Boundary Condition

C.2.5.1 Approximate Alternative

There are two possible ways to address the corner boundary condition issue. The more approximate of the two is to simply ignore the corner effects. This means that some sector of subgrade (a quadrant of a circle in most cases) beyond each corner of the mat that lies between the two straight edges of the mat that meet at that corner is left unaccounted for. Obviously, how important this is will vary from problem to problem depending on the total area of corner effects relative to areas included by edges.

The effect of this strategy was investigated to a limited extent as part of the initial development of the basic edge boundary condition that was presented above and originally in Colasanti and Horvath (2011). The two simple problems discussed in detail in Horvath and Colasanti (2011a) for the purposes of illustrating and assessing the Horvath-Colasanti/Reissner hybrid subgrade model were analyzed in *ANSYS* 11.0 using both a subgrade that extended explicitly beyond all portions of the mat[69] as well as one that ended at the edges of the mat with the special continuous k_{bc} boundary condition springs developed above. It was found that the error introduced by using the approximate boundary condition treatment of placing only continuous springs along the edges of the mat was never more than a few percent relative to the results obtained from the more exact treatment of modeling the entire subgrade explicitly beyond the limits of the mat.

The conclusion expressed in Colasanti and Horvath (2011) and reiterated here is that given all the other uncertainties that exist with subgrade modeling, even when using an advanced (multiple-parameter) model as in the case of the Horvath-Colasanti mechanical model, this small error is more than compensated by the ease of use in practice of the suggested edge-only boundary-condition springs as opposed to the effort of having to explicitly model the subgrade beyond the edge of the foundation.

Additional comments on this subject can be found in Chapter 6 as two of the three case histories presented in that chapter exhibited the more-common sagging (dish-shaped) settlement pattern. This allowed a comparison between measured and calculated (using the Horvath-Colasanti/Reissner hybrid subgrade model) results at corners. Obviously, the effect of neglecting the corner boundary condition are most noticeable at this location.

C.2.5.2 Rigorous Alternative

The alternative, theoretically rigorous approach is to develop a special subgrade boundary condition for use only at corners. Based on what was developed for the basic edge boundary condition, the most likely form of the outcome from this effort would be a single axial spring of stiffness K_{bc} placed beneath the corner at the lower spring level. This would be in addition to the continuous lines of edge springs placed at the same level as shown in Figure C.3c.

Two comments are in order before proceeding further. First, the proposed notation for this additional corner spring, K_{bc}, should not be confused with the use of that notation in Figure 6.1 and the figure on the back cover of this monograph. As noted previously in both this appendix and Chapter 6, the springs so labeled in these figures are actually the basic edge boundary condition springs, k_{bc}, discussed earlier in this appendix. The second comment is

---

[69] The subgrade was extended as necessary by trial and error beyond the edges of the mat until the calculated subgrade settlements were essentially zero.

that the dimensions of K_{bc}, force per length cubed, are different from the dimensions of k_{bc} which are force per length cubed per unit length.

It appears that the most fruitful approach for developing a solution for determining the stiffness of this special corner spring K_{bc} would be to work with cylindrical-polar (r, θ, z)[70] coordinates as opposed to the original Cartesian (x, y, z) coordinates. This is because corner boundary conditions will tend to involve accounting for some sector of a circle (most often a quadrant) in r-θ space that is not covered by the basic edge boundary conditions in x-y (really just x) space.

This effort will take some new theoretical development as the following steps appear to be necessary:

1. Both the Horvath-Colasanti (Equation C.1) and RSC (Equations C.2 and C.4) models that together comprise the Horvath-Colasanti/Reissner hybrid subgrade model used beneath the mat will need to be rederived for cylindrical-polar coordinates in r-θ-z space. Note that for the RSC, this means starting with the suites of partial-differential equations that define an elastic body expressed in cylindrical-polar coordinates and then making simplifying assumptions that are similar to those used to develop the original RSC solutions in Cartesian coordinates.

2. Using the results from Step 1, new equivalencies for the coefficients of these two models will need to be derived to replace those defined by Equations C.3a-c and C.5a-c. This is because the tension of the deformed/deformable membrane in the Horvath-Colasanti Model will have different units, force per radian, instead of force per unit length as in the original Cartesian coordinates.

3. The governing equation (C.6) of the Filonenko-Borodich Model that mechanically represents the subgrade behavior beyond the edge (actually the corner in this case) of the mat will need to be rederived for cylindrical-polar coordinates in r-θ-z space and then a new version of Equation C.7 that defines this model for the case where $p = 0$ will be required. Note that the equation that replaces Equation C.7 will be inherently more complex than Equation C.7 as settlement, w, will involve two spatial variables (r and θ) and not just one (x) as previously. However, it seems reasonable that as a first approximation at least the dependency of w on θ can be ignored[71] so that only the dependency of w on r needs be considered.

4. Using this new version of Equation C.7, a solution for $w = w(r)$ similar in concept to Equation C.8 would be obtained and then a solution process similar to that used for the basic edge boundary condition (i.e. Equations C.9 through C.12 inclusive) would follow. Note, however, that the end result would be for a boundary condition area of an entire circle so only a portion (one-quarter or $\pi/2$ radians corresponding to a quadrant corner in most cases) of the calculated spring stiffness would actually be used for K_{bc}.

---

[70] The variable θ here should not be confused with the same variable notation that was used earlier in this appendix to define the slope of the deformed/deformable tensioned membrane of the Horvath-Colasanti and Filonenko-Borodich models along the edge of the mat.

[71] Intuition suggests that for a given value of r the variation of w as a function of θ should be zero.

Appendix D

The Charles MT Method and Modifications (MTH Method)

D.1 ORIGINAL (MT) METHOD

D.1.1 Background

The MT Method (Charles 1996) was proposed by J. A. Charles who was then with the Building Research Establishment (BRE) in the U.K. The method was intended first and foremost to be used for forecasting the vertical normal stress distribution and concomitant depth to an equivalent rigid base for areally large earthworks and waste fills that applied an assumed uniform vertical stress increase to an existing ground surface. A secondary intended outcome was to forecast the settlement resulting from that stress increase.

In the context framed and used by Charles, 'depth to an equivalent rigid base' is perhaps better restated as what is sometimes called the 'significant stressed depth', i.e. the depth below which the stress increase caused by an applied stress can reasonably be assumed not to cause any additional settlement. This is a subtle but important point as the MT Method did not allow for any depth-wise change in subgrade material, certainly not a change with a concomitant significant increase in material stiffness that would be a de facto 'rigid base' regardless of stresses, e.g. a scenario of soil overlying bedrock. In fact, Charles assumed that the subgrade consisted of a homogeneous material that extended to some unlimited depth so that the forecast depth to rigid base was not influenced by a subgrade material change.

The underlying motivation for Charles' work was his opinion that the widely used 'one size fits all' rules of thumb for determining the significant stressed depth...the well-known $2B$ for a square loaded area of width B and $4B$ for an infinite strip of the same width such as found in Schmertmann (1978) and many other references...that were based on relative stress change alone, e.g. the depth at which the calculated stress change is some fixed percentage, often 10%, of the applied stress, was no longer defensible for use in practice. In particular, Charles opined that these rules of thumb were much too conservative for areally large applied loads that had effective 'rigid' bases at depths much less than $2B$. It is worth noting that Charles' position was in consonance with other research and published works of the same timeframe such as that of Burland and Burbidge (1985) and Berardi and Lancellotta (1991) that were focused on foundations and are discussed in Chapter 5. In particular, as discussed in Chapter 5, Burland and Burbidge found that in the limit for relatively large values of B that the depth to rigid base/significant stressed depth approached a value of only $0.3B$.

The underlying premise of Charles' work was that it was more relevant to evaluate the effect of stress changes at some depth relative to the vertical effective overburden stress that existed prior to load application. Qualitatively, the thinking was that a stress change that was small in magnitude relative to an existing overburden stress would be expected to have relatively little impact in terms of causing settlement regardless of the absolute magnitude or percentage of that stress change. So, a novel element of Charles' work is approaching the problem from a consideration of the absolute stresses as opposed to just stress change which has long been the cornerstone of many settlement-forecasting methodologies.

Not to detract from the fresh viewpoint that Charles brought to this overall problem but it is interesting to note that to some extent Charles' viewpoint was nothing new. From the earliest days of modern soil mechanics, the traditional 1-D compression analysis applied to oedometer test data on fine-grain soils has always evaluated the effects of a stress change relative to the absolute stress level to which the stress change (increase or decrease) is applied.

In any event, what is truly unique about Charles' work is the theoretical basis that Charles used for the primary assessment outcome of his methodology, forecasting the depth to rigid base. This can be seen in the name that Charles gave to his method which derives from the fact that its theoretical basis draws primarily and heavily on geomechanical concepts related to underground conduits and concomitant vertical arching that were collectively proposed by **M**arston and **T**erzaghi in the early years of the 20th century. Thus, the usual theoretical basis of the theory of elasticity that has been used historically for evaluating depth to rigid base/significant stressed depth played no role in Charles' theoretical development.

D.1.2 Overview

The central parameter and metric that was defined at the outset by Charles is the *load intensity ratio*, *n*. This dimensionless parameter is intended to reflect the scale effects of geotechnical applications in terms of both the:

- absolute size of loaded area in plan view and

- average vertical stress imposed on the subgrade by the loaded area

and their influence on the depth to rigid base. Specifically, this dimensionless parameter, which is used throughout Charles' paper, is the ratio of the average applied vertical stress, *q*, from the particular geotechnical application of interest (mat foundation in the case considered in this monograph but earthworks in Charles' original paper) to the pre-construction vertical effective overburden stress at a depth equal to what Charles called the *effective width*, *b**, of the applied stress[72].

As an aside, it can be seen that the *n* parameter effectively expresses the underlying conceptual premise of Charles' work that the controlling factor in terms of depth to rigid base/significant stressed depth was related to absolute vertical effective stress and not stress change.

In theory, *n* has a lower bound of zero (for an infinitely wide loaded area) but no upper bound and thus a semi-infinite range. However, from a practical perspective Charles showed that the variation in *n* is limited to approximately three orders of magnitude, even if one considers the entire gamut of foundation and geotechnical applications from tests on model-scale footings (where *n* can approach 100) to large earthworks (with *n* of the order of 0.1 or somewhat less).

Charles also found that the depth to an effective rigid base (*H* in the notation used in this monograph, z_d in the notation used by Charles) decreased with decreasing *n*, over a range of approximately 1½ orders of magnitude for the three-orders variation in *n*. This outcome is completely in consonance with the independent findings of Burland and Burbidge (1985) and Berardi and Lancellotta (1991). As a result, for relatively wide loaded areas such as

---

[72] *b** is a function of *L/B*, the actual length-to-width ratio of the applied stress in plan view. For a square loaded area (*L/B* = 1), *b** = *B*. In the limit where *L* → ∞, i.e. an infinite strip, *b** = 2*B*. Note that it is often assumed in practice that *L/B* ≥ 10 effectively constitutes *L* = ∞.

earthworks and mat foundations the depth to rigid base was considerably less than the common $2B$ criterion often universally assumed in practice as noted above. Charles indicated that for very wide loaded area with relatively low applied stresses (i.e. at the low end of the range of n) that depths to rigid base/significant stressed depths could easily be the $0.3B$ limit suggested by Burland and Burbidge (1985) and even somewhat less. So, despite the fact that Charles used a radically different analytical model compared to the combination of theory of elasticity plus observation of actual foundations that Burland and Burbidge used, the outcomes from these two bodies of work were completely in overall agreement both qualitatively and quantitatively.

As noted earlier in this appendix, another significant objective of the MT Method was settlement forecasting. The analytical methodology that Charles developed for this built upon and extended the same theoretical framework that he used for the depth-to-rigid-base forecast although it included some more-orthodox geomechanics elements of 1-D compression. Thus, the two primary outcomes of Charles' work, depth to rigid base/significant stressed depth and the settlement that occurs within that depth, can be viewed as one, integrated analytical methodology based on a single geomechanical behavioral model that is primarily based on concepts related to vertical arching.

D.1.3 Key Theoretical Elements

The theoretical aspects of the MT Method are relatively extensive and presented in detail in Charles (1996). Only a summary is presented here to provide context for the modifications and extensions made to the original MT Method by the writer when developing the writer's MTH Method.

As noted previously, the original MT Method has two distinct outcome components:

- forecasting the depth to rigid base/significant stressed depth and

- settlement.

The former can be calculated and used alone if desired but is necessary if the latter is also desired.

Overall, both components are based on a 1-D compression model for the subgrade beneath a loaded area. Clearly, the assumption of 1-D compression becomes a better approximation of reality as the width of the loaded area becomes relatively large which was the application of primary interest to Charles.

The specific 1-D model that Charles developed and used for forecasting the depth to rigid base is stress-based as would be expected. However, as noted earlier in this appendix it is arguably unconventional in its philosophy for this application as the calculation of the vertical effective stress, σ'_v, is based on Marston's concept of vertical arching over underground conduits. Elements of Terzaghi's contributions to vertical arching (colloquially referred to as the 'trapdoor problem') are used as well.

The primary soil properties used in the behavioral model postulated by Charles are the Mohr-Coulomb strength parameter, ϕ, and the coefficient of lateral earth pressure, K_h (referred to as K in Charles' paper). These parameters are used to estimate the vertical shearing resistance that develops within the assumed vertical prism of soil within which the assumed geomechanism of vertical arching is occurring. It is this vertical shearing resistance that Charles assumed was resisting the applied surface load, q. With regard to the latter

174

parameter, K_h/K, Charles assumed that it was not only equal to the normally consolidated at-rest value, K_{oNC}, but could be evaluated using the common empirical expression[73]:

$$K_{oNC} = 1 - \sin\phi \,.$$

(D.1)

As a result, the only input soil properties required for forecasting the depth to rigid base are the effective soil unit weight[74], γ_{eff}, and ϕ, both of which were assumed to be single-valued for a particular site and application. When combined with the parameter n that had been calculated previously, a unique calculated result for the depth to rigid base can be obtained.

To extend Charles' analytical methodology to include forecasting settlements, the only additional soil property required is the drained constrained modulus, D. As with ϕ and γ_{eff}, Charles assumed that D was single-valued for any given site and application.

The central element of Charles' model for estimating settlement is the following assumption as to how vertical strain, ε_v, and change in vertical effective stress, $\Delta\sigma'_v$, are calculated as a function of depth, z, beneath the applied load, q:

$$\varepsilon_v = \frac{\Delta\sigma'_v}{D} = \frac{\sigma'_v - \gamma_{eff}z}{D} \,.$$

(D.2)

For cases involving the drained behavior of fine-grain soils, D can be determined directly from oedometer data using the following relationship (Lambe and Whitman 1969):

$$D = \frac{\sigma'_{v-avg}}{0.435 \cdot CR}$$

(D.3)

where CR = *compression ratio* (interpreted from the oedometer test data with ε_v vs. $\log_{10}\sigma'_v$ plotting) and σ'_{v-avg} = the average vertical effective stress[75]. Two of the case histories discussed in Chapter 6 used this approach.

---

[73] As the writer discussed in detail in Horvath (2004), for a period of time there was active discussion and concomitant lack of consensus in the geotechnical engineering research community as to whether the value of ϕ used in Equation D.1 should be that corresponding to peak-strength, ϕ_{peak}, or constant-volume (critical-state), ϕ_{cv}, conditions. Subsequently, the consensus fell on the latter. However, Charles did not acknowledge no less take a position in this discussion. Rather, he implied that whatever general value for ϕ was selected for a particular application would be used in Equation D.1.

[74] This is the soil unit weight that produces an effective vertical overburden stress in a given application. Charles only noted the two limiting cases in his paper. One was that the entire subgrade depth was partially saturated in which case $\gamma_{eff} = \gamma_t$, the total/bulk/damp/wet unit weight. The other case was that the entire subgrade depth was saturated and below the groundwater table in which case $\gamma_{eff} = \gamma_b = \gamma'$, the buoyant unit weight. Charles did not address the case where the groundwater table could lie at some intermediate depth within the overall depth of interest in a given application so that $\gamma_b < \gamma_{eff} < \gamma_t$.

[75] Note that Equation D.3 applies only if σ'_{v-avg} is within the virgin loading range. If it is within the reload range, then the *recompression ratio, RR*, must be substituted for CR in this equation. Note also that if e vs. $\log_{10}\sigma'_v$ plotting is used for the oedometer data, then the appropriate relationships between the *compression index, C_c*, and CR and the *recompression index, C_r*, and RR must be used before using Equation D.3. In either case, the situation becomes more complex if the loading covers both the recompression and virgin loading ranges as some average of the relevant compression parameters must be used. Charles did not explicitly address any of these potential variations as he simply assumed that a single value of D was known for the subgrade.

If the subgrade stiffness is evaluated during the site characterization investigation in terms of Young's modulus, E, then Poisson's ratio must also be assumed and the following theoretical relationship used to convert to D:

$$D = \frac{E(1-v)}{(1+v)(1-2v)}.$$ **(D.4)**

If the subgrade stiffness is evaluated during the site characterization investigation in terms of shear modulus, G, then Poisson's ratio must also be assumed and either one of the following theoretical relationships used to convert to D:

$$E = 2G(1+v) \ then \ use \ Equation \ D.4 \ or$$ **(D.5)**

$$D = \frac{2G(1-v)}{(1-2v)}.$$ **(D.6)**

Note that in both cases, the values of E or G must correlate with the operative stress range of interest, i.e. similar in concept to the use of $\sigma'_{v\text{-}avg}$ in Equation D.3.

D.2 MODIFICATIONS AND EXTENSIONS (MTH METHOD)

D.2.1 Background

In 2006, the writer conducted research to modify the MT Method to allow for greater generality and analytical flexibility in the overall methodology. The goal of this work was to enhance the viability of the method for use with hybrid subgrade models and their application to mat foundations. The culmination of this work is referred to as the MTH Method.

The overall attraction of the MT Method is that the calculated outcomes inherently include an estimate of the depth to rigid base. Up to that point in time, the writer had been using the Fraser-Wardle Method that is discussed in Chapter 5 for converting a layered system (Figure 5.1b) to an equivalent single-layer system (Figure 5.1c) during the process illustrated and discussed in Chapter 5. As noted that chapter, a significant downside of using the Fraser-Wardle Method is that it requires an a priori estimate of the depth to rigid base. As illustrated using the case histories presented in Chapter 6, some geologic profiles lend themselves to making a reasonable visual estimate of the depth to rigid base while others do not. Thus, the analytically inclusive nature inherent in the MT Method with respect to the depth-to-rigid-base issue was a significant attraction for applying it to SSI applications in general and mat foundations in particular even though the method was originally developed with a very different non-foundation, geotechnical application in mind.

D.2.2 Objectives

The focus of the writer's research related to the MT Method was the subgrade profile that could be analyzed. Specifically, the primary objective was to get away from the relatively significant restrictions of the original MT Method that essentially treated the subgrade as a single layer as it required single values for the necessary input parameters (ϕ and γ_{eff} as a minimum, D as well if settlements were to be calculated). Thus, allowing for a layered system in terms of these input parameters was a top priority.

An additional objective of the writer's research was to incorporate an improved algorithm for embedment effects in the overall methodology. The original MT Method was crafted with earthworks in mind so it was assumed that the applied stress, q, was at the original ground surface. With mat foundations, embedment is the rule, not the exception, and can be significant in many cases. Although Charles made suggestions for dealing with embedment, they were subjective in nature and judged by the writer to be overall unsatisfactory for general use.

D.2.3 Changes Implemented in MTH Method

The primary change implemented by the writer in the MTH Method was to allow for depth-wise variation in the key problem parameters/soil properties (γ_{eff}, ϕ, D, ν) by modeling the actual subgrade (Figure 5.1a) as a system of m number of arbitrary layers (Figure 5.1b) where only a given layer has to have single-value parameters/soil properties. The calculation of vertical strain (Equation D.2) is then performed separately for each layer.

Additional changes incorporated into the MTH Method were:

- for each artificial layer, allow for different values of D for reloading and virgin loading depending on the relationship of the calculated stress level, σ'_v, relative to the yield stress, σ'_{vm}, for that layer which was also input; and

- allow for K_o values other than K_{oNC} as was assumed in the original MT Method (Equation D.1) using the well-known empirical relationship with *Overconsolidation Ratio, OCR*, that is also an input variable:

$$K_0 = K_{oNC} \cdot OCR^{(1-\sin\phi)} = (1 - \sin\phi) \cdot OCR^{(1-\sin\phi)} \,. \tag{D.7}$$

The equivalent averages of both the drained constrained modulus, D^*, and Poisson's ratio, ν^*, for the overall system shown in Figure 5.1c are obtained using the following relationships based on an assumption of strain-weighting over m number of layers:

$$D^* = \frac{\sum_{i=1}^{m}(D_i \varepsilon_{vi})}{\sum_{i=1}^{m} \varepsilon_{vi}} \tag{D.8}$$

$$\nu^* = \frac{\sum_{i=1}^{m}(\nu_i \varepsilon_{vi})}{\sum_{i=1}^{m} \varepsilon_{vi}} \,. \tag{D.9}$$

Finally, the equivalent, average, single-value Young's modulus, E, for the system shown in Figure 5.1c is calculated by rearranging Equation D.4 to solve for E using the results from Equations D.8 and D.9. The equivalent shear modulus, G, for the system shown in Figure 5.1c is calculated by rearranging Equation D.6 to solve for G using the results from Equations D.8 and D.9.

Note that the averages shown in Equations D.8 and D.9 are for m artificial layers between foundation level (not the ground surface) and the effective rigid base at a depth H below foundation level. The latter depth is unknown initially and always needs to be determined on a site- and application-specific basis using the same logic as in the original MT

Method, i.e. as the depth at which the stress change, $\Delta\sigma'_v$, as defined in Equation D.2 equals zero. For the case histories presented in Chapter 6, this was done on a trial-and-error basis incorporated within a proprietary FORTRAN computer code named *MTH* that was developed by the writer in 2009 (Horvath 2009b). There may well be other solutions strategies and it may be possible to perform the necessary tasks in generic spreadsheet software.

D.2.4 Potential Changes to MTH Method

It has been over a decade since the writer developed the MTH Method. During that timeframe, there have been a number of technological advances in several areas that are relevant to the method. Most of these advances involve the combination of wider commercial availability of the seismic piezocone, sCPTu, and the development of numerous empirical correlations between measured parameters in sCPTu and piezocone, CPTu, soundings and fundamental soil properties and other geotechnical parameters than can readily be incorporated even into the site characterization process for routine projects in practice. As a collective result, there are a number of areas where the MTH Method could be improved upon while still staying within the basic theoretical framework proposed by Charles in 1996.

Some of the specific potential areas of improvement that the writer has identified to date are:

- Direct use of K_o determined from the site-characterization process as opposed to calculating this parameter based on input values of ϕ and *OCR*.

- Allowing for greater flexibility in the choice of input ϕ values, e.g. the use of ϕ_{peak}, not ϕ_{cv}, given the fact that subgrade strains are relatively small and unlikely to be large enough to mobilize the constant-volume/critical-state strength state. This flexibility is possible because direct input of K_o de-links ϕ which was used to both estimate K_o (which required ϕ_{cv} in Equation D.7) and soil strength in both the original MT and writer's MTH methods.

- Use of shear modulus G rather than D as the input stiffness parameter for each artificial layer. This could make use of G_{max}/G_o data generated from sCPTu soundings combined with the Fahey-Carter modulus-degradation model (Fahey and Carter 1993) but based on strain, not stress for which the Fahey-Carter Model was originally developed. Niazi (2014) showed that the concept of shear-modulus degradation based on strain worked well for deep-foundation applications so there is no reason why it would not work well in SSI applications. Guidance as to what degradation limits to use for this purpose could be developed from data shown in Berardi and Lancellotta (1991) and possibly other references.

Finally, the use of generic spreadsheet software such as *Excel* for all calculations should be investigated.

This page intentionally left blank.

REFERENCES

ACI Committee 336 (1988). "Suggested Analysis and Design Procedures for Combined Footings and Mats". *ACI Structural Journal*, Vol. 85, No. 3, pp. 304-324.

ACI Committee 336 (1989). Closure to "Suggested Analysis and Design Procedures for Combined Footings and Mats". *ACI Structural Journal*, Vol. 86, No. 1, pp. 113-116.

Banavalkar, P. V. (1995). "Mat Foundation and Its Interaction with the Superstructure". In *Design and Performance of Mat Foundations; State-of-the-Art Review*, American Concrete Institute, pp. 13-49.

Banavalkar, P. V. and Ulrich, Jr., E. J. (1982). "Vertical Movement: A Design and Construction Challenge Provided by Tall Buildings with Deep Excavations". Proceedings of Ground/Expo 82, Houston, TX.

Banavalkar, P. V. and Ulrich, Jr., E. J. (1984). "RepublicBank Center: Structural and Geotechnical Features". Proceedings of the International Conference on Tall Buildings, Singapore.

Berardi, R. and Lancellotta, R. (1991). "Stiffness of Granular Soils from Field Performance". *Géotechnique*, Vol. 41, No. 1, pp. 149-157.

Biot, M. A. (1937). "Bending of an Infinite Beam on an Elastic Foundation". *Journal of Applied Mechanics/Transactions - American Society of Mechanical Engineers*, Vol. 4, No. 1, pp. A1-A7.

Bosson, G. (1939). "The Flexure of an Infinite Elastic Strip on an Elastic Foundation". *Philosophical Magazine*, Vol. 27, pp. 37-50.

Bowles, J. E. (1988). *Foundation Analysis and Design*, 4th edition. McGraw-Hill.

Brown, P. T. (1969). "Numerical Analyses of Uniformly Loaded Circular Rafts on Deep Elastic Foundations". Géotechnique, Vol. 19, No. 3, pp. 399-404.

Brown, P. T. (1974). "Influence of Soil Inhomogeneity on Raft Behaviour". *Soils and Foundations*, Vol. 14, No. 1, pp. 61-70.

Burland, J. B., Broms, B. B., and de Mello, V. F. B. (1977). "Behaviour of Foundations and Structures". Proceedings of the Ninth International Conference on Soil Mechanics and Foundation Engineering, Tokyo, Japan, Vol. 2, pp. 495-546.

Burland, J. B. and Burbidge, M. C. (1985). "Settlement of Foundations on Sand and Gravel". *Proceedings of the Institution of Civil Engineers*, Part 1, Vol. 78, pp. 1325-1381.

Chambers, R. E. (1984). *Structural Plastics Design Manual - Volume I*. ASCE Manuals and Reports on Engineering Practice No. 63, American Society of Civil Engineers.

Charles, J. A. (1996). "The Depth of Influence of Loaded Areas". *Géotechnique*, Vol. 46, No. 1, pp. 51-61.

Colasanti, R. J. and Horvath, J. S. (2011). "A Practical Subgrade Model for Improved Soil-Structure Interaction Analysis: Software Implementation". *Practice Periodical on Structural Design and Construction*, Vol, 15, No. 4.

Dawkins, W. P. (1982). *User's Guide: Computer Program for Analysis of Beam-Column Structures with Nonlinear Supports (CBEAMC)*. Instruction Report K-82-6, U.S. Army Engineer Waterways Experiment Station, Vicksburg, MS.

DeSimone, S. V. and Gould, J. P. (1972). "Performance of Two Mat Foundations on Boston Blue Clay". Proceedings of the Specialty Conference on Performance of Earth and Earth-Supported Structures, American Society of Civil Engineers, Vol. 1, Part 2, pp. 953-980.

Douglas, R. A. (1995). Discussion of "Modelling of Geosynthetic-Reinforced Engineered Granular Fill on Soft Soil" by S. K. Shukla and S. Chandra, *Geosynthetics International*, Vol. 2, No. 4, pp. 771-775.

Fahey, M. and Carter, J. P. (1993). "A Finite Element Study of the Pressuremeter Test in Sand Using a Nonlinear Elastic Plastic Model". *Canadian Geotechnical Journal*, Vol. 30, No. 2, pp. 348-362.

Focht, Jr., J. A., Khan, F. R., and Gemeinhardt, J. P. (1978). "Performance of One Shell Plaza". *Journal of the Geotechnical Engineering Division*, Vol. 104, No. 5, pp. 593-608.

Fraser, R. A. and Wardle, L. S. (1976). "Numerical Analysis of Rectangular Rafts on Layered Foundations". *Géotechnique*, Vol. 26, No. 4, pp. 613-630.

Gazetas, G. (1981). "Ultimate Behavior of Continuous Footings in Tensionless Contact with a Three-Parameter Soil". *Journal of Structural Mechanics*, Vol. 9, No. 3, pp. 339-362.

Ghosh, C. and Madhav, M. R. (1994a). "Settlement Response of a Reinforced Shallow Earth Bed". *Geotextiles and Geomembranes*, Vol. 13, No. 10, pp. 643-656.

Ghosh, C. and Madhav, M. R. (1994b). "Reinforced Granular Fill-Soft Soil System: Confinement Effect". *Geotextiles and Geomembranes*, Vol. 13, No. 11, pp. 727-741.

Ghosh, C. and Madhav, M. R. (1994c). "Reinforced Granular Fill-Soft Soil System: Membrane Effect". *Geotextiles and Geomembranes*, Vol. 13, No. 11, pp. 743-759.

Haber-Schaim, I. (1973). *Geomechanics*. Privately published in Israel.

Haliburton, T. A. (1971). *Soil Structure Interaction: Numerical Analysis of Beams and Beam-Columns*. Technical Publication 14, Oklahoma State University, School of Civil Engineering, Stillwater, OK.

Harr, M. E. (1966). *Foundations of Theoretical Soil Mechanics*. McGraw-Hill.

Harr, M. E., Davidson, J. L., Ho, D.-M., Pombo, L. E., Ramaswamy, S. V., and Rosner, J. C. (1969). "Euler Beams on a Two Parameter Foundation Model". *Journal of the Soil Mechanics and Foundations Division*, Vol 95, No. 4, pp. 933-948.

Hemsley, J. A. (2000). "Developments in Raft Analysis and Design". In *Design Applications of Raft Foundations*, Thomas Telford, pp. 487-605.

Hetényi, M. (1946). *Beams on Elastic Foundation*. The University of Michigan Press, Ann Arbor, MI.

Hetényi, M. (1950). "A General Solution for the Bending of Beams on an Elastic Foundation of Arbitrary Continuity". *Journal of Applied Physics*, Vol. 21, pp. 55-58.

Horikoshi, K. and Randolph, M. F. (1997). "On the Definition of Raft-Soil Stiffness Ratio for Rectangular Rafts". *Géotechnique*, Vol. 47, No. 5, pp. 1055-1061.

Horvath, J. S. (1979). *A Study of Analytical Methods for Determining the Response of Mat Foundations to Static Loads.* Dissertation presented to the Polytechnic Institute of New York, Department of Civil Engineering, Brooklyn, NY in partial fulfillment of the requirements for the degree of Doctor of Philosophy.

Horvath, J. S. (1983a). "New Subgrade Model Applied to Mat Foundations". *Journal of Geotechnical Engineering*, Vol. 109, No. 12, pp. 1567-1587.

Horvath, J. S. (1983b). "Modulus of Subgrade Reaction: New Perspective". *Journal of Geotechnical Engineering*, Vol. 109, No. 12, pp. 1591-1596.

Horvath, J. S. (1984a). "Simplified Elastic Continuum Applied to the Laterally Loaded Pile Problem - Part I: Theory". *Laterally Loaded Deep Foundations: Analysis and Performance - Special Technical Publication 835*, American Society for Testing and Materials, pp. 112-121.

Horvath, J. S. (1984b). Errata in "New Subgrade Model Applied to Mat Foundations". *Journal of Geotechnical Engineering*, Vol. 110, No. 8, p. 1171.

Horvath, J. S. (1984c). Errata in "Modulus of Subgrade Reaction: New Perspective". *Journal of Geotechnical Engineering*, Vol. 110, No. 8, p. 1171.

Horvath, J. S. (1988a). *Numerical Analysis of Beams and Beam-Columns with Linear and Non-Linear Spring Supports Using Finite Differences.* Research Report CE/GE-88-1, Manhattan College, School of Engineering, Civil Engineering Department, Geotechnical Engineering Program, Bronx, NY.

Horvath, J. S. (1988b). *Further Evaluation of the Coefficient, or Modulus, of Subgrade Reaction, k, Using an Extension of Reissner's Simplified Elastic Continuum Concept.* Research Report CE/GE-88-2, Manhattan College, School of Engineering, Civil Engineering Department, Geotechnical Engineering Program, Bronx, NY.

Horvath, J. S. (1988c). *Determination of the Coefficient of Subgrade Reaction, k, for Beams.* Research Report CE/GE-88-3, Manhattan College, School of Engineering, Civil Engineering Department, Geotechnical Engineering Program, Bronx, NY.

Horvath, J. S. (1988d). *Historical Review and Critique of Mathematical Models for Plate- and Beam-Type Foundation Element Subgrades.* Research Report CE/GE-88-4, Manhattan College, School of Engineering, Civil Engineering Department, Geotechnical Engineering Program, Bronx, NY.

Horvath, J. S. (1989a). Discussion of "Suggested Analysis and Design Procedures for Combined Footings and Mats" by ACI Committee 336, *ACI Structural Journal*, Vol. 86, No. 1, pp. 112-113.

Horvath, J. S. (1989b). *Base Pressure and Earth Pressure Measurements; Nürnberg Subway.* Research Report CE/GE-89-1, Manhattan College, School of Engineering, Civil Engineering Department, Geotechnical Engineering Program, Bronx, NY.

Horvath, J. S. (1989c). "Subgrade Models for Soil-Structure Interaction". Proceedings of the Foundation Engineering Congress, American Society of Civil Engineers, pp. 599-612.

Horvath, J. S. (1990). *A Theoretical Study of the Effect of Anchors on the Performance of Shallow Foundations under Vertical Loads.* Research Report CE/GE-90-3, Manhattan College, School of Engineering, Civil Engineering Department, Geotechnical Engineering Program, Bronx, NY.

Horvath, J. S. (1992a). Discussion of "Modified Vlasov Model for Beams on Elastic Foundations" by C. V. G. Vallabhan and Y. C. Das. *Journal of Geotechnical Engineering*, Vol. 118, No. 9, pp. 1482-1484.

Horvath, J. S. (1992b). "Mat Foundation Analysis: A Review and Critique Based on Case Histories". Preprint paper, American Concrete Institute Spring 1992 Convention.

Horvath, J. S. (1993a). "Analysis of Anchored Foundations". Preprint paper 93-0078, 72nd Annual Meeting, Transportation Research Board, Washington, DC.

Horvath, J. S. (1993b). "Cut-and-Cover Tunnel Subgrade Modeling". Preprint paper 93-0087, 72nd Annual Meeting, Transportation Research Board, Washington, DC.

Horvath, J. S. (1993c). "Subgrade Modeling for Soil-Structure Interaction Analysis of Horizontal Foundation Elements". Paper presented at a special joint meeting of the Structural and Geotechnical groups of the American Society of Civil Engineers Metropolitan Section, New York, NY.

Horvath, J. S. (1993d). *Subgrade Modeling for Soil-Structure Interaction Analysis of Horizontal Foundation Elements.* Research Report CE/GE-93-1, Manhattan College, School of Engineering, Civil Engineering Department, Geotechnical Engineering Program, Bronx, NY.

Horvath, J. S. (1993e). "Beam-Column-Analogy Model for Soil-Structure Interaction Analysis". *Journal of Geotechnical Engineering*, Vol. 119, No. 2, pp. 358-364.

Horvath, J. S. (1993f). Errata in "Beam-Column-Analogy Model for Soil-Structure Interaction Analysis". *Journal of Geotechnical Engineering*, Vol. 119, No. 7, p. 1183.

Horvath, J. S. (1993g). "Cut-and-Cover Tunnel Subgrade Modeling". *Transportation Research Record 1415*, Transportation Research Board, Washington, DC.

Horvath, J. S. (1994). Discussion of "The Effect of Prestressing on the Settlement Characteristics of Geosynthetic-Reinforced Soil" by S. K. Shukla and S. Chandra, *Geotextiles and Geomembranes*, Vol. 13, No. 11, pp. 761.

Horvath, J. S. (1995). "Mat Foundation Analysis: A Review and Critique Based on Case Histories". In *Design and Performance of Mat Foundations; State-of-the-Art Review*, American Concrete Institute, pp. 117-160.

Horvath, J. S. (2002). *Soil-Structure Interaction Research Project; Basic SSI Concepts and Applications Overview*. Research Report CGT-2002-2, Manhattan College, School of Engineering, Center for Geotechnology, Bronx, NY.

Horvath, J. S. (2004). *Integrated Site Characterization and Foundation Analysis Research Project; A Technical Note re Effect of K_{onc} Assumption on Site-Characterization Algorithm for Coarse-Grain Soil*. Research Report CGT-2004-2, Manhattan College, School of Engineering, Center for Geotechnology, Bronx, NY.

Horvath, J. S. (2009a). *SSIH: A Computer Program for Soil-Structure Interaction Analysis of Horizontal Foundation Elements; User Instructions and Guidelines*. Published by John S. Horvath, Ph.D., P.E.

Horvath, J. S. (2009b). *MTH: A Computer Program for Calculating the Equivalent Depth to Rigid Base and Pseudo-Homogenous/Isotropic Single-Layer Elastic Parameters of a Layered System Using an Extended Version of the Charles MT Model; User Instructions and Guidelines*. Published by John S. Horvath, Ph.D., P.E.

Horvath, J. S. (2011). *Soil-Structure Interaction Research Project; A Practical Subgrade Model for Improved Soil-Structure Interaction Analysis: Parameter Assessment*. Research Report CEEN/GE-2011-1, Manhattan College, School of Engineering, Civil and Environmental Engineering Department, Geotechnical Engineering Program, Bronx, NY.

Horvath, J. S. (2016a). "*Addendum #1 to New Developments in Site Characterization and Pile-Resistance Calculation: A White Paper*". Published by John S. Horvath Consulting Engineer, Scarsdale, NY.

Horvath, J. S. (2016b). "*Addendum #2 to New Developments in Site Characterization and Pile-Resistance Calculation: A White Paper*". Published by John S. Horvath Consulting Engineer, Scarsdale, NY.

Horvath, J. S. and Colasanti, R. J. (2011a). "A Practical Subgrade Model for Improved Soil-Structure Interaction Analysis: Model Development". *International Journal of Geomechanics*, Vol. 11, No. 1, pp. 59-64.

Horvath, J. S. and Colasanti, R. J. (2011b). "New Hybrid Subgrade Model for Soil-Structure Interaction Analysis: Foundation and Geosynthetics Applications". Proceedings of Geo-Frontiers 2011, ASCE Geo-Institute/IFAI/GMA/NAGS.

Horvilleur, J. F. and Patel, V. (1995). "Mat Foundation Design - A Soil-Structure Interaction Problem". In *Design and Performance of Mat Foundations; State-of-the-Art Review*, American Concrete Institute, pp. 51-94.

Jones, R. and Xenophontos, J. (1976). "On the Vlasov and Kerr Foundation Models". *Acta Mechanica*, Vol. 25, Issue 1-2, pp. 45-49.

Kamesovara Rao, N. S. V., Das, Y. C., and Anandakrishnan, M. (1971). "Variational Approach to Beams on Elastic Foundations". *Journal of the Engineering Mechanics Division*, Vol. 97, No. 2, pp271-294.

Kerr, A. D. (1961). "Viscoelastic Winkler Foundation with Shear Interactions". *Journal of the Engineering Mechanics Division*, Vol. 87, No. 3, pp. 13-30.

Kerr, A. D. (1964). "Elastic and Viscoelastic Foundation Models". *Journal of Applied Mechanics*, Vol. 31/*Transactions of the American Society of Mechanical Engineers*, Vol. 86, pp. 491-498.

Kerr, A. D. (1965). "A Study of a New Foundation Model". *Acta Mechanica*, Vol. 1, No. 2, pp. 135-147.

Kerr, A. D. (1966). *A Study of a New Foundation Model*. Research Report 186, U.S. Army Cold Regions Research Laboratory, Hanover, NH.

Kerr, A. D. (1985). "On the Determination of Foundation Model Parameters". *Journal of Geotechnical Engineering*, Vol. 111, No. 11, pp.1334-1340.

Kerr, A. D. and Rhines, W. J. (1967). *A Further Study of Elastic Foundation Models*. Report S-67-1, New York University, School of Engineering and Science, Department of Aeronautics and Astronautics, Bronx, NY.

Kerr, W. C. and Saxena, S. K. (1977). "Role of Cracking in Deformations of Shallow Foundations". Preprint paper 3000, 1977 Fall Convention, American Society of Civil Engineers.

Kneifati, M. C. (1985). "Analysis of Plates on a Kerr Foundation Model". *Journal of Engineering Mechanics*, Vol. 111, No. 11, pp. 1325-1342.

Lambe, T. W. and Whitman, R. V. (1969). *Soil Mechanics*. Wiley.

Liao, S. S. C. (1991). *Estimating the Coefficient of Subgrade Reaction for Tunnel Design*. Final draft report of an internal research project sponsored by Parsons Brinckerhoff, Inc., New York, NY.

Liao, S. S. C. (1995). "Estimating the Coefficient of Subgrade Reaction for Plane Strain Conditions". *Geotechnical Engineering*, Vol. 113, No. 3, pp. 166-181.

Loof, H. W. (1962). "The Theory of the Coupled Spring Foundation". *Heron*, Delft Institute of Technology, Delft, The Netherlands, pp. 63-79. [in Dutch]

Loof, H. W. (1965). "The Theory of the Coupled Spring Foundation". *Heron*, Delft Institute of Technology, Delft, The Netherlands, pp. 29-49.

Mayne, P. W. (2005). "Unexpected but Foreseeable Mat Settlements on Piedmont Residuum". *International Journal of Geoengineering Case Histories*, Vol. 1, No. 1, pp. 5-17.

Mayne, P. W. (2006). "In-Situ Test Calibrations for Evaluating Soil Parameters". Proceedings of the Second International Workshop on Characterisation and Engineering Properties of Natural Soil, Volume 3 - Chapter 2. Taylor and Francis.

Mayne, P. W. (2007). *Cone Penetration Testing*. National Cooperative Highway Research Program Synthesis 368, Transportation Research Board, Washington, DC.

Mayne, P. W. and Poulos, H. G. (1999). "Approximate Displacement Influence Factors for Elastic Shallow Foundations". *Journal of Geotechnical and Geoenvironmental Engineering*, Vol. 125, No. 6, pp. 453-460.

Nair, K. and Chang, C.-Y. (1973). *Flexible Pavement Design and Management - Materials Characterization*. National Cooperative Highway Research Program Report 140, Highway Research Board, Washington, DC.

Niazi, F. S. (2014). *Static Axial Pile Foundation Response Using Seismic Piezocone Data*. Dissertation presented to Georgia Institute of Technology, College of Engineering, School of Civil and Environmental Engineering, Atlanta, GA in partial fulfillment of the requirements for the degree of Doctor of Philosophy.

Pister, K. S. and Williams, M. L. (1960). "Bending of Plates on a Viscoelastic Foundation". *Journal of the Engineering Mechanics Division*, Vol. 86, No. EM5, pp. 31-44.

Poulos, H. G. and Davis, E. H. (1974). *Elastic Solutions for Soil and Rock Mechanics*. Wiley.

Reissner, E. (1958). "A Note on Deflections of Plates on a Viscoelastic Foundation". *Journal of Applied Mechanics*, Vol. 25/*Transactions of the American Society of Mechanical Engineers*, Vol. 80, pp. 144-155.

Reissner, E. (1967). "Note on the Formulation of the Problem of the Plate on an Elastic Foundation". *Acta Mechanica*, Vol. 4, pp. 88-91.

Rhines, W. J. (1965). *Foundation Models for Continuously Supported Structures*. Dissertation presented to New York University, School of Engineering and Science, Department of Aeronautics and Astronautics, Bronx, NY in partial fulfillment of the requirements for the degree of Doctor of Philosophy.

Rhines, W. J. (1966). "A Study of the Reissner Foundation Model". Proceedings of the Fifth National Congress on Applied Mechanics.

Rhines, W. J. (1967). *A Study of the Reissner Foundation Model*. Report S-67-5, The National Aeronautics and Space Administration.

Roark, R. J. and Young, W. C. (1975). *Formulas for Stress and Strain*, 5th edition. McGraw-Hill.

Robertson, P. K. and Cabal, K. L. (2015). *Guide to Cone Penetration Testing for Penetration Testing*, 6th edition. Gregg Drilling & Testing, Inc., Signal Hill, CA.

186

Schiffman, R. L., Whitman, R. V., and Jordan, J. C. (1970). "Settlement Problem Oriented Computer Language". *Journal of the Soil Mechanics and Foundation Engineering Division*, Vol. 96, No. SM2, pp. 649-669.

Schmertmann, J. H. (1978). *Guidelines for Cone Penetration Test Performance and Design.* Report FHWA-TS-78-209, U.S Department of Transportation, Federal Highway Administration, Washington, DC.

Scott, R. F. (1981). *Foundation Analysis.* Prentice-Hall.

Shukla, S. K. and Chandra, S. (1994a). "The Effect of Prestressing on the Settlement Characteristics of Geosynthetic-Reinforced Soil". *Geotextiles and Geomembranes*, Vol. 13, No. 8, pp. 531-543.

Shukla, S. K. and Chandra, S. (1994b). "A Study of Settlement Response of a Geosynthetic-Reinforced Compressible Fill-Soft Soil System". *Geotextiles and Geomembranes*, Vol. 13, No. 9, pp. 627-639.

Shukla, S. K. and Chandra, S. (1994c). "A Generalized Mechanical Model for Geosynthetic-Reinforced Foundation Soil". *Geotextiles and Geomembranes*, Vol. 13, No. 12, pp. 813-825.

Shukla, S. K. and Chandra, S. (1995). "Modelling of Geosynthetic-Reinforced Engineered Granular Fill on Soft Soil". *Geosynthetics International*, Vol. 2, No. 3, pp. 603-618.

Terzaghi, K. (1943). *Theoretical Soil Mechanics.* Wiley.

Terzaghi, K. (1955). "Evaluation of Coefficients of Subgrade Reaction". *Géotechnique*, Vol. 5, No. 4, pp. 297-326.

The Institution of Structural Engineers (1989). *Soil-Structure Interaction; The Real Behaviour of Structures.* U.K.

Timoshenko, S. P. and Gere, J. M. (1961). *Theory of Elastic Stability*, 2nd edition. McGraw-Hill.

Timoshenko, S. P. and Gere, J. M. (1972). *Mechanics of Materials.* Van Nostrand.

Timoshenko, S. P. and Woinowsky-Krieger, S. (1959). *Theory of Plates and Shells*, 2nd edition. McGraw-Hill.

Tomlinson, M. J. (1986). *Foundation Design and Construction*, 5th edition. Wiley.

Ulrich, Jr., E. J. (1991). "Subgrade Reaction in Mat Foundation Design". *Concrete International*, April, pp. 41-50.

Ulrich, Jr., E. J. and Jacob, K. P. (1979). "Houston Skyscrapers: Designed to Move but Not Too Much". *Soundings*, McClelland Engineers, 2nd Quarter issue.

Vallabhan, C. V. G. and Das, Y. C. (1991). "Modified Vlasov Model for Beams on Elastic Foundations". *Journal of Geotechnical Engineering*, Vol. 117, No. 6, pp. 956-966.

Vesic, A. B. and Johnson, W. H. (1963). "Model Studies of Beams Resting on a Silt Subgrade". *Journal of the Soil Mechanics and Foundations Division*, Vol. 89, No. SM1, pp. 1-31.

Vesic, A. S. and Saxena, S. K. (1970). *Analysis of Structural Behavior of AASHO Road Test Rigid Pavements*. National Cooperative Highway Research Program Report 97, Highway Research Board, Washington, DC.

Vlasov, V. Z. and Leont'ev, N. N. (1960). *Beams, Plates and Shells on Elastic Foundations* (*Balki, Plity i Obolochki na Upragom Osnovanii*). Gosudartstvennoe Izdatel'stvo, Fiziko-Matematicheskoi Literatury, Moscow, U.S.S.R.; translated from Russian and published in 1966 by the Israel Program for Scientific Translations, Jerusalem, Israel, in cooperation with The National Aeronautics and Space Administration and The National Science Foundation, Washington, DC.

Westergaard, H.M. (1926) "Stresses in Concrete Pavements Computed by Theoretical Analysis". *Public Roads*, Vol. 7, pp. 25-35.

Yin, J. H. (1997a). "Modelling Geosynthetic-Reinforced Granular Fills Over Soft Soils". *Geosynthetics International*, Vol. 4, No. 2, pp. 165-185.

Yin, J. H. (1997b). "A Nonlinear Model of Geosynthetic-Reinforced Granular Fill Over Soft Soil". *Geosynthetics International*, Vol. 4, No. 5, pp. 523-537.

This page intentionally left blank.

www.ingramcontent.com/pod-product-compliance
Lightning Source LLC
Chambersburg PA
CBHW050239220326
41598CB00047B/7453

* 9 7 8 1 7 3 2 0 9 5 3 1 1 *